INVITATION TO COMBINATORIAL TOPOLOGY

INVITATION TO COMBINATORIAL TOPOLOGY

MAURICE FRÉCHET AND KY FAN

Translated from the French, with some notes,
by Howard W. Eves

DOVER PUBLICATIONS, INC.
MINEOLA, NEW YORK

Bibliographical Note

This Dover edition, first published in 2003, is an unabridged republication of the work published in 1967 as *Initiation to Combinatorial Topology* by Prindle, Weber & Schmidt, Incorporated, Boston, as volume seven of its Complementary Series in Mathematics, under the consulting editorship of Howard W. Eves.

Library of Congress Cataloging-in-Publication Data

Fréchet, Maurice, 1878-
    [Introduction à la topologie combinatoire. English]
    Invitation to combinatorial topology / Maurice Fréchet and Ky Fan ; translated from the French, with some notes by Howard W. Eves.
       p. cm.
    Originally published: Initiation to combinatorial topology. Boston : Prindle, Weber & Schmidt, in series: The Prindle, Weber & Schmidt complementary series in mathematics; v. 7.
    Includes bibliographical references and index.
    ISBN 0-486-42786-2 (pbk.)
    1. Combinatorial topology. I. Fan, Ky. II. Eves, Howard Whitley, 1911- III. Title.

QA611 .F34713 2003
514'.22—dc21

2002035173

Manufactured in the United States of America
Dover Publications, Inc., 31 East 2nd Street, Mineola, N.Y. 11501

Foreword

This book may, perchance, fall into the hands of a reader curious about mathematics, but who, before beginning to read our work, wishes to have an idea, expressed in nontechnical language and in a few lines, as to *what is topology* (also called *analysis situs*).[1] H. Poincaré ([19],[2] pages 57–61) has already given satisfaction to this reader in an intuitive and particularly striking way:

"Geometers usually distinguish two kinds of geometry, the first of which they qualify as metric and the second as projective. Metric geometry is based on the notion of distance; two figures are there regarded as equivalent when they are 'congruent' in the sense that mathematicians give to this word. Projective geometry is based on the notion of straight line; in order for two figures considered there to be equivalent, it is not necessary that they be congruent; it suffices that one can pass from one to the other by a projective transformation, that is, that one be the perspective of the other. This second body of study has often been called qualitative geometry, and in fact it is if one opposes it to the first; it is clear that measure and quantity play a less important role. This is not entirely so, however. The fact that a line is straight is not purely qualitative; one cannot assure himself that a line is straight without making measurements, or without sliding on this line an instrument called a straightedge, which is a kind of instrument of measure.

"But it is a third geometry from which quantity is completely excluded and which is purely qualitative; this is *analysis situs*. In this discipline, two figures are equivalent whenever one can pass from one to the other by a continuous deformation; whatever else the law of this deformation may be, it must be continuous. Thus a circle is equivalent to an ellipse or even to an arbitrary closed curve, but it is not equivalent to a straight line segment since this segment is not closed. A sphere is

---

[1] The name *topology* has only recently and gradually replaced the expression *analysis situs*.

[2] Numbers in brackets refer to the bibliographical list on pages 73 and 74.

equivalent to any convex surface; it is not equivalent to a torus since there is a hole in a torus and in a sphere there is not. Imagine an arbitrary design and a copy of this same design executed by an unskilled draftsman; the properties are altered, the straight lines drawn by an inexperienced hand have suffered unfortunate deviations and contain awkward bends. From the point of view of metric geometry, and even of projective geometry, the two figures are not equivalent; on the contrary, from the point of view of *analysis situs*, they are.

"*Analysis situs* is a very important science for the geometer; it leads to a sequence of theorems as well linked as those of Euclid; and it is on this set of propositions that Riemann has built one of the most remarkable and most abstract theories of pure analysis. I will cite two of these theorems in order to clarify their nature: (1) Two plane closed curves cut each other in an even number of points. (2) If a polyhedron is convex, that is, if one cannot draw a closed curve on its surface without dividing it in two,[1] the number of edges is equal to that of the vertices plus that of the faces diminished by two, and this remains true when the faces and the edges of the polyhedron are curved.

"And here is what makes this *analysis situs* interesting to us; it is that geometric intuition really intervenes there. When, in a theorem of metric geometry, one appeals to this intuition, it is because it is impossible to study the metric properties of a figure as abstractions of its qualitative properties, that is, of those which are the proper business of *analysis situs*. It has often been said that geometry is the art of reasoning correctly from badly drawn figures. This is not a capricious statement; it is a truth that merits reflection. But what is a badly drawn figure? It is what might be executed by the unskilled draftsman spoken of earlier; he alters the properties more or less grossly; his straight lines have disquieting zigzags; his circles show awkward bumps. But this does not matter; this will by no means bother the geometer; this will not prevent him from reasoning correctly.

"But the inexperienced artist must not represent a closed curve as an open curve, three lines which intersect in a common point by three lines which have no common point, or a pierced surface by a surface without

---

[1] AUTHORS' NOTE: Poincaré's text, intentionally written in a form adapted to his aim of popularization, is not rigorous from the point of view of mathematics. We should replace the words "that is" by "or more generally."

a hole. Because then one would no longer be able to avail himself of his figure and reasoning would become impossible. Intuition will not be impeded by the flaws of drawing which concern only metric or projective geometry; it becomes impossible as soon as these flaws relate to *analysis situs*.

"This very simple observation shows us the real role of geometric intuition; it is to assist this intuition that the geometer needs to draw figures, or at the very least to represent them mentally. Now, if he ignores the metric or projective properties of these figures, if he adheres only to their purely qualitative properties, it will be then and only then that geometric intuition really intervenes. Not that I wish to say that metric geometry rests on pure logic and that true intuition never intervenes there, but these are intuitions of another kind, analogous to those which play the essential role in arithmetic and in algebra."

Translator's Preface

It has been a delight constructing an English translation of the little classic, *Initiation à la Topologie Combinatoire*, of Maurice Fréchet and Ky Fan. The original French work, which appeared in 1946, has been allowed to go out of print, and a Spanish translation published in South America a few years ago is hardly accessible here in the United States.

The French work had its origin in a course given by Professor Fréchet in 1935. Later, when Professor Fréchet felt that an elementary intuitive introduction to combinatorial topology, somewhat along the lines of his earlier course, might be of interest to a wider audience outside the university, he found himself too occupied with other matters to undertake the project personally. It was at this point that the project was fortunately saved by Dr. Ky Fan, who volunteered to carry out the task. This he did in a most masterful way, producing a veritable gem of elementary exposition. Anyone having a small acquaintance with high school geometry can read this work with comprehension. There are only three short sections which assume a somewhat higher mathematical knowledge, and, following the plan of the French work, these are here printed in smaller type and can be skipped without harm to the understanding of the main body of the text.

In translating the work, it was felt that perhaps a few notes by the translator might be welcome. It is hoped that these notes, though sometimes a trifle more demanding than the text proper, are still reasonably within the spirit of the original work. They supply proofs of some of the unproved statements of the original text, point out some extensions and applications of concepts of the work, and engage in some historical comment; they are given only for the reader who would like a little more. It is generally recommended that these notes be read after the completion of the text proper, for there are occasions where a note, appearing in reference to some point of the text, requires subsequent material of the text for its understanding.

The main attempt here, of course, has been to convert the original text into English, and in this way to try to capture for the English-reading public some of the expository and pedagogical genius of Maurice Fréchet and Ky Fan.

In conclusion, it is a pleasure to acknowledge the cooperation and encouragement given by Professor Ky Fan, who at this writing is located at the University of California at Santa Barbara.

HOWARD W. EVES

Contents

Foreword   v
Translator's Preface   ix

CHAPTER ONE.   TOPOLOGICAL GENERALITIES

 1   Qualitative Geometric Properties   1
 2   Coloring Geographical Maps   2
 3   The Problem of Neighboring Regions   5
 4   Topology, India-Rubber Geometry   6
 5   Homeomorphism   7
 6   Topology, Continuous Geometry   11
 7   Comparison of Elementary Geometry, Projective Geom-
       etry, and Topology   12
 8   Relative Topological Properties   14
 9   Set Topology and Combinatorial Topology   17
10   The Development of Topology   19

CHAPTER TWO.   TOPOLOGICAL NOTIONS ABOUT SURFACES

11   Descartes' Theorem   21
12   An Application of Descartes' Theorem   25
13   Characteristic of a Surface   27
14   Unilateral Surfaces   29
15   Orientability and Nonorientability   31
16   Topological Polygons   35
17   Construction of Closed Orientable Surfaces from Polygons
       by Identifying Their Sides   36
18   Construction of Closed Nonorientable Surfaces from
       Polygons by Identifying Their Sides   40
19   Topological Definition of a Closed Surface   45

CHAPTER THREE. TOPOLOGICAL CLASSIFICATION OF CLOSED SURFACES

20 The Principal Problem in the Topology of Surfaces   49
21 Planar Polygonal Schema and Symbolic Representation of a Polyhedron   50
22 Elementary Operations   53
23 Use of Normal Forms of Polyhedra   55
24 Reduction to Normal Form: I   56
25 Reduction to Normal Form: II   59
26 Characteristic and Orientability   64
27 The Principal Theorem of the Topology of Closed Surfaces   67
28 Application to the Geometric Theory of Functions   69
29 Genus and Connection Number of Closed Orientable Surfaces   69

Bibliography   73

TRANSLATOR'S NOTES   75

Index   120

CHAPTER ONE

Topological Generalities

1. QUALITATIVE GEOMETRIC PROPERTIES

The properties of figures explicitly stated in elementary geometry are, for the most part, metric properties—that is to say, properties depending on size or measure. Among such properties are, for example, the congruence of two triangles, the equality of two angles, the condition for a quadrilateral to be a square, etc. But certain properties of figures are quite independent of size or measure and are not separately encountered in elementary geometry. On the one hand, for example, consider the interior of a circle, the interior of an ellipse, and the interior of a square, and on the other hand, a ring-shaped region formed by two concentric circles in the plane [a].[1] All these figures clearly have different metric properties. Nevertheless, intuition tells us that there are some properties which are common to the first three figures but which the last does not possess. For example, the first three figures have this property in common: the region bounded by any simple closed polygonal line lying in the interior of the figure also belongs entirely to the interior of the figure. It is clear that the circular ring does not have this property. One can thus say that there are certain *qualitative properties* that the interior of a circle, the interior of an ellipse, and the interior of a square possess in common, and which the circular ring does not possess.

Consider, further, a circular circumference in the plane. This circumference divides the rest of the plane into two parts, such that two points in a common part are always able to be joined by a polygonal line

---

[1] Lower-case boldface letters in brackets refer to the translator's notes on pages 75–119.

in the plane which does not cut the circumference, whereas every polygonal line in the plane joining two points belonging to the two different parts cuts the circumference. But one can suitably modify the metric nature of the circumference without altering this property: if we replace the circumference by an ellipse or by a simple closed planar polygonal line, this property still holds. In a general way, all these figures are particular cases of what is called a *closed Jordan curve* (see further, page 13). It is the celebrated Jordan theorem that states that *every closed Jordan curve in a plane divides the rest of the plane into two parts* (in the precise sense described above). We thus have a qualitative property of the plane. In spite of the intuitive evidence of Jordan's theorem, a proof of the theorem is not easy [b]. (There are more than a score of proofs of this theorem. For example, see C. JORDAN [12], pages 92–100; A. DENJOY [5]; DE KERÉKJÁRTÓ [14], page 21, and [15]; E. SCHMIDT [21]; F. SEVERI [23], pages 24–35.)

In order better to reveal the existence of qualitative properties of figures, we shall look at some problems which are, moreover, interesting in themselves.

2. COLORING GEOGRAPHICAL MAPS[1]

Cayley in 1878 attributed to De Morgan the announcement of the *four-color theorem* [c]. According to this theorem, if an arbitrary geographical map represents countries (or political subdivisions) by their boundaries [d], then no more than four colors are sufficient to color the map so that any two countries touching each other along a common boundary have different colors. (Moreover, Frederick Guthrie asserted in 1880 that his brother Francis Guthrie had given a proof of this theorem as early as thirty years before.) For some time various different but false proofs were held to be sound. Actually, no one as yet has proved this theorem. But it has been shown[2] that five colors always suffice [e], and it is easy to see that three colors are not in general sufficient. Consider, for example, a map of seven regions A, B, C, D, E,

[1] Cf. A. SAINTE-LAGUË [20]; A. ERRERA [6]; D. HILBERT and S. COHN-VOSSEN [11], pages 294–300.

[2] Cf. P. J. HEAWOOD [9]; A. ERRERA [6], page 38.

F, G situated as shown in Fig. 1. One easily ascertains that four colors are needed to color these seven regions. [It is interesting to note that, on the map of France, the département of Seine-et-Oise and the neighboring départements (Seine, Oise, Seine-et-Marne, Loiret, Eure-et-Loire, Eure) have a mutual relationship of the type pictured in Fig. 1.] One can even construct a map of only four regions for which fewer than four colors will not be sufficient. Fig. 2 shows a map of four regions in which every two are adjacent; it therefore requires four colors to color the regions.

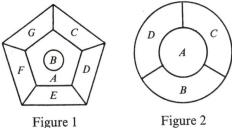

Figure 1   Figure 2

In generalizing the four-color problem, one can pose the following: Given a surface, determine the minimum number of colors sufficient to color all possible maps on this surface so that any two regions touching along a curve have different colors (but two regions having only a finite number of points in common may have the same color). This minimum number is called the *chromatic number* of the surface.

In the case of a plane or a sphere, the chromatic number appears in all likelihood to be 4. But this is a conjecture that is very difficult to establish [**f**]. Nevertheless, the problem has been resolved for several kinds of surfaces.

Take, for example, the *torus*, which is the surface obtained by rotating a circle about a nonintersecting line in its plane (Fig. 3). It has been shown[1] that the chromatic number of the torus is 7. We shall content ourselves here with showing that fewer than seven colors are not sufficient. To accomplish this, it suffices to prove that it is possible to construct seven regions on the torus such that each pair of these regions touch each other along a curve. Now this is possible. Cut the torus along a generating circle, which will convert the surface into a deformed

---

[1] P. J. HEAWOOD [9]; or else A. ERRERA [6], page 58.

cylinder. Straighten this and cut the cylinder along one of its altitudes; a rectangle will be obtained (Fig. 4).[1] It follows that the regions on the

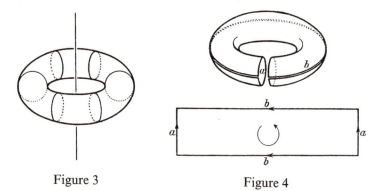

Figure 3   Figure 4

torus can be represented on the rectangle, care being taken to identify the parallel sides of the rectangle. In Fig. 5, a division of the torus into seven regions is represented on a rectangle, the four regions G of the rectangle forming only one region on the torus. Each pair of these seven regions touch each other along a curve [**g**].

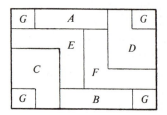

Figure 5

Note that an analogous problem cannot be posed concerning spatial domains.

It was Frederick Guthrie who first made the observation that, for any positive integer n, one can construct n domains in space such that each pair of them touch each other along a surface, from which one concludes that one needs n colors to color the domains. A simple example is obtained in the following way. Cut (Fig. 6) a parallelepiped by a plane

---

[1] The curved arrow marked in the rectangle will be used later (page 37).

parallel to the base, obtaining in this way two superposed parallele-pipeds. Divide these two parallelepipeds respectively into n partial parallelepipeds, numbered from 1 to n, one by planes parallel to one direction of the lateral faces and the other by planes parallel to the other direction of the lateral faces. Consider as domain number k ($k = 1, 2,$

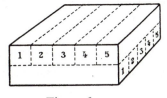

Figure 6

$\cdots$, n) the pair of partial parallelepipeds numbered k. It is seen that each pair of these n domains touch each other along a surface; in order to color them, then, one must have n colors.

3. THE PROBLEM OF NEIGHBORING REGIONS

This problem[1] is intimately related to the problem of coloring geographic maps. The problem is to determine, for a given surface, the maximum number of regions [h] which can be constructed on the surface such that each pair of these regions touch each other along a curve.[2] It is seen that the chromatic number of a surface is at least equal to this maximum number of regions adjacent in pairs, for to color the regions adjacent in pairs it is necessary to have as many colors as there are regions. We have seen (Fig. 2) that the maximum number of regions adjacent in pairs for the plane is at least 4. It can be shown[3] [i] that it is exactly 4. In the case of the torus, the two numbers are equal; the common value is 7.[4]

---

[1] Cf. L. HEFFTER [10].

[2] As in the problem of coloring geographic maps, these regions do not have to cover the entire surface.

[3] A. ERRERA [6], page 58.

[4] Since the chromatic number of the torus is 7, the maximum number of regions adjacent in pairs on the torus ≤ 7. But, according to Fig. 5, this maximum number of regions adjacent in pairs ≥ 7. It follows that this maximum number is 7.

Let us mention the further problem of determining the maximum number of points on a given surface that can be joined in pairs by curves drawn on the surface with no two of the curves intersecting one another. Consider, for example, five points on a plane. There are ten pairs of these points. It is impossible to draw on the plane ten non-intersecting curves such that each pair of the five points constitutes extremities of one of the curves. In other words, if one succeeds in drawing nine curves, not intersecting one another, joining nine pairs of these points, every curve joining the tenth pair must cut at least one of the other nine curves (Fig. 7). The sought maximum number of points

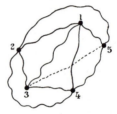

Figure 7

for the plane is thus less than 5. In fact, it is 4 [i]. And, similarly, one can also prove (see HILBERT and COHN-VOSSEN [11], page 295) that, for every surface, the maximum number in question is equal to the maximum number of regions adjacent in pairs [i]. In particular, there then exist on the torus seven points (but there do not exist eight) which can be joined in pairs by curves drawn on the torus with no two of these curves intersecting each other.

4. TOPOLOGY, INDIA-RUBBER GEOMETRY

In the preceding problems, one deals with neither size nor measure; one is concerned only with certain qualitative properties. And, moreover, these qualitative properties have a common character that is important to clarify.

Imagine a torus made of a kind of elastic matter like India-rubber, which we are able to distort as we wish in a continuous fashion without tearing. When the torus is deformed, without tearing or overlapping, into another surface, the answer to any one of the preceding problems for the resulting surface will be the same as for the torus. One can, for example, find seven regions adjacent in pairs on the torus; it is clear that the same can be said for the resulting surface.

Similarly, imagine a figure made of India-rubber and deform it without tearing or overlapping. Certain properties of the figure are changed by the deformation, but there are other properties that remain unchanged. Those properties that remain invariant under the deformation are the so-called *topological properties* of the figure. It is topological properties that one studies in *topology*.

For example, if the surface of a circle is made of India-rubber, one can deform it, without tearing or overlapping, into the surface of an ellipse or the surface of a square, etc. The properties common to the surface of a circle, to the surface of an ellipse, to the surface of a square, and, more generally, to all figures obtained by deforming the surface of a circle without tearing or overlapping, are studied in topology.

Thus, vaguely and roughly speaking, we can say that topology is the geometry of India-rubber figures. The concept of homeomorphism will allow us to make this precise.

5. HOMEOMORPHISM

When one distorts, without tearing or overlapping, a figure made of India-rubber, one notices that there are two remarkable relationships between the initial figure and that obtained by the deformation: to each point of the one figure corresponds one and only one point of the other, and to two neighboring points of the one correspond two neighboring points of the other. These two relationships can be expressed in a mathematical form that we are going to make precise, and they are, in fact, the only relationships that matter, as far as we are concerned, in a "deformation, without tearing or overlapping, of an India-rubber figure."

Let E and F be two figures (or two sets of points), distinct or not. Suppose that to each point of E there corresponds a definite point of F, the same point of F perhaps corresponding to several points of E, and, moreover, each point of F corresponding to at least one point of E. We shall say of each such correspondence that it determines a (*unique*) *transformation*[1] f of the figure E onto the figure F. If b denotes the point of F corresponding to a point a of E, we write $b = f(a)$, and say that b is the *image*, or the *transform*, of the point a, or that f *transforms* a into b.

Now let f be a transformation of E onto F. In general, to two distinct points of E may correspond the same point of F. In the particular case where to different points a_1 and a_2 of E there always correspond two different points $f(a_1)$ and $f(a_2)$ of F, we say that f is a *biunique transformation*, or a *biunique correspondence* [j], between the points of E and those of F. There exists, then, for each point b of F, one and only one point $a = g(b)$ of E such that $f(a) = b$, and the correspondence g is called the *inverse transformation* of f, a particular transformation of set F onto set E. The transformation g, which is the inverse of f, is often designated by f^{-1}.

A transformation f of E onto F is *continuous* when points of E which are close to a point a of E are transformed into points of F which are close to the point $f(a)$. More precisely, we say that f is *continuous at a point a of E* if, for each number $\varepsilon > 0$ there exists a number $\eta > 0$ such that, for every point x of E at a distance less than η from a, the point $f(x)$ in F is at a distance less than ε from $f(a)$. If f is continuous at each point of E, we say f is a *continuous transformation of E onto F*.

A biunique transformation f of E onto F is said to be *bicontinuous* if it is continuous from E onto F and if, moreover, the inverse transformation f^{-1} is continuous from F onto E.

A biunique and bicontinuous transformation is called a *homeomorphism* or a *topological transformation*. If we allow ourselves to speak intuitively, we can say that *a homeomorphism between two figures (or two sets of points) is a correspondence such that to each point of either one of the two figures corresponds one and only one point of the other and to two neighboring points of either correspond two neighboring points of the other.*

---

[1] More precisely, we should say (*unique*) *point transformation*. Otherwise one can envisage (as, for example, the transformation by polar reciprocation) a correspondence f between a figure E composed of points and a figure F composed of lines, f making a line of F correspond to each point of E. This, then, is not a point transformation. But, as we shall deal only with point transformations, we shall, for simplicity, suppress the word "point."

We spoke of the "deformation, without tearing or overlapping, of an India-rubber figure" in order to introduce the idea of homeomorphism. Actually, the "deformation without tearing or overlapping" is only a very particular process for the concrete realization of a homeomorphism. It is possible to achieve a homeomorphism by other processes. For example, let us cut a torus along the circumference of a generating circle

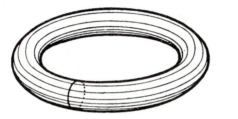

Figure 8

to obtain a surface in the form of a tube with two ends. From this surface we make a knot (Fig. 8), and then rejoin the two ends in such a manner that points which coincided on the torus coincide on the final surface. One then sees that there exists a biunique and bicontinuous correspondence between the torus and the surface of the knot. We have passed from the first surface to the second by means of cutting, deforming, and pasting. This very simple example shows that we cannot consider a homeomorphism as the exact mathematical translation of "deformation without tearing or overlapping," to which another mathematical concept, that of *isotopy*, better applies, and of which we shall say more later (page 16). We used "deformation without tearing or overlapping" only to emphasize the principal property of a homeomorphism—that of being a biunique and bicontinuous transformation [k].

Two figures (or two sets of points) are said to be *homeomorphic* if one can pass from one to the other by a homeomorphism, that is, if there exists a biunique and bicontinuous transformation between them. If E and F are homeomorphic, and similarly F and G, it is clear that E and G are homeomorphic.

Examples of homeomorphic figures are numerous. Two figures that are superposable by displacement are homeomorphic. It follows that to show that two figures E and G are homeomorphic, it suffices to prove that E is homeomorphic to a figure F superposable on G by displacement.

As we have just seen, a torus and the surface of a knot (Fig. 8) are two homeomorphic figures [l]. The surface of a sphere S and that of a tetrahedron T are homeomorphic: by replacing, if necessary, the sphere S by an equal sphere S', we can suppose that the center of sphere S' is in the interior of the tetrahedron; we can then project the one surface onto the other from the center of the sphere. One easily sees that this projection establishes a homeomorphism between the two surfaces S' and T, whence, according to the preceding, S and T are also homeomorphic.

The plane and the surface of a sphere from which one point has been removed are homeomorphic. By the same method of displacement as used above on the sphere S, suppose that the plane is tangent to the sphere at the point diametrically opposite the removed point. Then project the surface of the sphere (except for the removed point) on the plane by taking the removed point as center of projection (Fig. 9). One sees that this is a biunique transformation. It is also bicontinuous: to two neighboring points (different from the removed point) on the sphere there correspond two neighboring points on the plane, and conversely. This homeomorphism is known under the name of *stereographic projection*.

A point and a line clearly are two nonhomeomorphic figures, for no biunique correspondence is possible between them. Similarly, a line segment and the surface of a square are not homeomorphic. Intuition seems to tell us that this is visually evident, for one feels that the surface of a square should be infinitely richer in points than a segment. But this is false; contrary to our expectation, one can establish a biunique correspondence between the points of the surface of a square and those of a segment [**m**]. It will accordingly not be as simple as thought at first sight to prove that a segment and the surface of a square are not homeomorphic. What is necessary to show is that a biunique correspondence between a segment and the surface of a square cannot be bicontinuous.

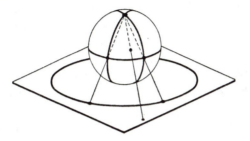

Figure 9

In topology one often encounters apparently evident facts which, in reality, are difficult to prove[1] or are even incorrect. It is difficult, for example, to imagine a set in space which is homeomorphic to a segment and which is such that its projection on a plane fills the entire surface of a square. Such a set exists, however.

6. TOPOLOGY, CONTINUOUS GEOMETRY

Topology, or *analysis situs*, is a modern branch of geometry which, as we have seen, does not bring in the notions of size or measure, but only that of continuity. It concerns itself, then, only with qualitative properties of figures.

More precisely, one can define the aim of topology as follows. A property of a set is said to be *topological* if it can be expressed by means of the concept of continuity. A topological property of a set is called a *topological invariant* if it is preserved under all homeomorphisms. *Topology is the study of topological properties and, especially, topological invariants of figures.*

---

[1] For example, the Jordan theorem pointed out on page 2.

It is important to note that a topological property of a set is not necessarily a topological invariant, in which case the property is said to be *relative*. In other words, two homeomorphic sets E and F can have certain different topological properties, as when it is a matter not of properties of E and F alone, but of properties concerning E and F and the spaces which contain them. In Section 8 we shall return in more detail to relative topological properties.

The concept of homeomorphism plays in topology the same role that congruence plays in elementary geometry. Two congruent figures have the same properties in elementary geometry. Similarly, since the chief properties studied in topology are the topological invariants, and since these are common to any two homeomorphic sets, two such sets ought to be regarded in topology as essentially nondifferent or, more accurately, as equivalent.

Thus two homeomorphic sets are *topologically equivalent*. The relation of homeomorphism between sets can be considered as an equivalence relation, for the following three conditions of equivalence are easily verified: (1) every set is homeomorphic to itself; (2) if E is homeomorphic to F, then F is homeomorphic to E; (3) if E is homeomorphic to F and F is homeomorphic to G, then E is also homeomorphic to G. Consequently, sets can be divided into pairwise disjoint[1] *topological classes* such that two homeomorphic sets belong to a common topological class and two nonhomeomorphic sets to two distinct topological classes [**n**].

7. COMPARISON OF ELEMENTARY GEOMETRY, PROJECTIVE GEOMETRY, AND TOPOLOGY

In order better to understand the aim of topology, let us present a comparison of elementary geometry, projective geometry, and topology.

To do this, let us first say a little about projective geometry.[2] It was Poncelet who made projective geometry a self-contained doctrine. The

---

[1] Two sets of elements are said to be *disjoint* if they have no element in common.

[2] Cf., for example, L. GODEAUX [8], pages 62–102.

underlying idea in the work of this geometer can be summarized as follows. Let E be a plane figure in space and let O be a point outside the plane of the figure. Project the figure E from point O—that is, draw the lines determined by O and the points of E; we then obtain a new figure F in space. Cut F by a plane distinct from that of E and not containing O; the section gives us a new plane figure G.[1] From known properties of figure E, one can deduce some properties of figure G. Thus Poncelet utilized two operations: projection and section. He termed projective figures any two figures which can each be obtained one from the other by a finite number of projections and sections. The aim of projective geometry is the study of properties common to two projective figures.

Thus, though ellipses, parabolas, and hyperbolas have different properties in elementary geometry, they have the same properties in projective geometry, since these are projective figures of one another. Therefore, from the point of view of projective geometry, one does not distinguish the three types of conics.

Similarly, from the topological point of view, any two homeomorphic sets have the principal topological properties in common (at least they have the same topological invariants). Thus a circle, an ellipse, and a simple closed polygonal line have, in the eyes of the topologist, no essential difference. They are all *closed Jordan curves*—that is, sets homeomorphic to a circle.

Given a geometric figure, its properties, from the point of view of elementary geometry, are the properties preserved by all displacements of the figure. If one next considers its properties from the point of view of projective geometry, one still retains the properties preserved by projection and section. Finally, if one adopts the topological point of view, it remains principally[2] for us to study the topological invariants. Thus the properties of the figure become rarer and rarer as we pass from elementary geometry to projective geometry and then to topology. But, though it is true that the topological properties are less numerous than those of elementary geometry or those of projective geometry, it is no

---

[1] We thus obtain figure G from figure E by two operations: projection and section. But one sometimes says, for brevity, that figure E is projected from center O onto figure G.

[2] We say "principally" because in topology one can also consider topological properties which are not topological invariants.

less true that the topological properties are the most fundamental properties of figures. One can compare elementary geometry to a man dressed in many-colored clothes, projective geometry to a naked body, and topology to the human skeleton [o].

8. RELATIVE TOPOLOGICAL PROPERTIES

Although two homeomorphic figures have the same topological invariants, they can have different topological properties in those properties that concern their relations with the spaces which contain them.

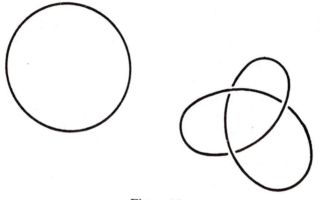

Figure 10

Return, for example, to the torus and the surface of a knot (Fig. 8), of which we have spoken on page 9. These are two homeomorphic figures. but one can show that there exists no homeomorphism of the entire space onto itself such that the torus is transformed into the surface of a knot. As another example, consider in space a circle and a twisted curve forming a knot[1] (Fig. 10); one can easily establish a biunique and bicontinuous correspondence between the points of these two figures,

---

[1] The *Hoppe curve*, represented in rectangular coordinates by the parametric equations

$$x = \cos t \,(3 \cos t + 2), \quad y = 5 \cos t \sin t, \quad z = \sin t \,(25 \cos^2 t - 1),$$

assumes the form of a knot.

which are thus homeomorphic, but one cannot pass from one to the other by a homeomorphism of the entire space onto itself.

More generally, we say that two figures have (topologically) the same *position in space* if one can pass from one to the other by a homeomorphism of the entire space onto itself. One sees that this condition implies that the two figures are homeomorphic. The above examples show that the converse is not true: two homeomorphic figures may not have the same position in space.

A topological property of a figure is said to be *relative* if it depends on the position of the figure in space.

Let us cite another example. It was Gauss who, by the use of a certain integral, made correspond to each pair of nonintersecting closed Jordan curves in space a whole number called their *linking number* (in French: nombre d'enlacements; in German: Verschlingungszahl). This number is zero if and only if the two curves are *not linked* in the sense that one of them bounds a region of a surface[1] not cutting the other. One sees that the property of a pair of closed Jordan curves (with no point in common) of having linking number zero is certainly a topological property, but it is a relative property of the pair. In fact, two such closed Jordan curves are always homeomorphic; the linking number for a pair is determined only by the position of the pair of curves in space.

Let us point out here, following L. ANTOINE [3], that there are, from this point of view, three kinds of homeomorphic figures E and F in the plane:

Case 1: There exists a homeomorphism of the entire plane onto itself carrying E onto F.

Case 2: No homeomorphism of the entire plane onto itself transforms E onto F; but one can pass from E to F by a homeomorphism between a conveniently chosen neighborhood of E and a conveniently chosen neighborhood of F (understanding by "neighborhood of a figure E" any plane set V containing E and such that for each point a of E there exists a circular disk with center a and positive radius belonging entirely to V).

Case 3: One cannot pass from E to F by a homeomorphism between a

---

[1] More precisely, it is necessary to say *orientable* surface. For the significance of the word "orientable," see page 31.

neighborhood of E and a neighborhood of F; but there exists a homeo-morphism between E and F.

Each of these three cases actually exists. We have, for example, case 1 if E and F are two closed Jordan curves (see ANTOINE [3], pages 232–241). We have cases 2 and 3, respectively, if E and F are two figures like those represented in Fig. 11 and in Fig. 12 (see ANTOINE [3], page 251).

In the first case, the two figures have, by their very definition, the same position in the plane. On the contrary, in the last two cases, the two

Figure 11   Figure 12

figures do not have the same position in the plane and they are homeo-morphic figures having different relative topological properties.

To the relative topological properties are attached two further im-portant concepts—that of *homotopy* and that of *isotopy*—which tie in with observations made on page 9. Consider, first of all, two particular examples. It is clear that one can deform the surface of a sphere into that of an ellipsoid in such a way that for each intermediate stage in the course of the deformation the surface remains homeomorphic to the sphere. On the other hand, in order to establish a homeomorphism between the torus and the surface of a knot (page 9), we had first to cut the torus along a generating circumference, and then, after having made a knot, we glued the two ends together. Consider, in the course of these operations, the intermediate stage where the surface has the form of a tube with two ends. Because of the presence of these two ends, this sur-face is not homeomorphic to the torus. And one can moreover prove that it is impossible to deform the torus in a continuous way into the surface of a knot.

This very simple observation leads, then, to the following definition. We say that figure E is *isotopic* to figure F in space if one can pass from E to F by a continuous family of homeomorphisms, so that at each instant the intermediate figure is homeomorphic to E. More generally,

E is said to be *homotopic* to F in space if one can pass from E to F by a continuous family of single-valued continuous transformations.

Instead of studying isotopy or homotopy in space, one can also consider, more generally, isotopy or homotopy of E to F in a set of points containing E and F. Here is the precise definition:

Consider three point sets E, F, G such that the first two belong to the third, and further consider the interval $0 \leq t \leq 1$ (t can, if one wishes, be regarded as the measure of time). Suppose that to each point a of E and to each number t in the interval $0 \leq t \leq 1$ there corresponds a point of G, which we designate by $f_t(a)$, such that the following conditions are satisfied: (1) for $t = 0$, we have $f_0(a) = a$ for all a in E; for $t = 1$, the points $f_1(a)$ form the set F as a ranges over E; (2) the point $f_t(a)$ depends continuously on t and a as t and a range, respectively, over the interval $0 \leq t \leq 1$ and the set E. Under these conditions, we shall say that the family of transformations f_t ($0 \leq t \leq 1$) is a *homotopy* of E to F on the set G, or, equivalently, that E is *homotopic* to F in G.

If one further knows that, for each value t_0 of t, f_{t_0} is a homeomorphism between E and the set of points $f_{t_0}(a)$ (a ranging over the set E), we shall say that the family of homeomorphisms f_t is an *isotopy* between E and F on the set G, or, equivalently, E is *isotopic* to F in G. Isotopy is thus a particular case of homotopy.

If two sets E and F are isotopic in a set G containing E and F, they are necessarily homeomorphic. But the converse is not true—the surface of a torus and that of a knot are two homeomorphic figures, but they are not isotopic in space and consequently not in any set containing them.

9. SET TOPOLOGY AND COMBINATORIAL TOPOLOGY

We have seen which properties of figures are studied in topology. But, more fundamentally, which figures are studied in topology?

One can take *a priori*, for the subject of topology, all kinds of point sets, as one does in a branch of topology called *set topology*—that is to say, topology based on the theory of sets. Since the important work of G. Cantor, the creator of the theory of sets, one is convinced that point sets of a very general nature give rise to interesting topological questions. Some methods of the theory of sets are often needed to treat these questions. Since the figures studied in set topology are extremely general sets, it is natural that the results obtained in this way are frequently far from intuitive and even sometimes in contradiction with intuition. (For example, the fact that there exist in the plane three different domains having identical boundaries[1] is certainly far from intuitive [**p**].) This

---

[1] See DE KERÉKJÁRTÓ [14], page 120.

leads to a finical desire to correct our often very clumsy intuition and to emphasize pathological situations. Delicacy of reasoning is a characteristic feature of set topology. On the other hand, in spite of its great generality, set topology finds many applications in mathematical analysis. Thanks to the efforts of a large number of geometers, most of whom are contemporary, set topology has achieved a very important place in the mathematical sciences.

But in order to make possible the application of certain fruitful and especially convenient methods, there is an interest in limiting the subject of topology to simple and usual geometric figures, such as curves and surfaces, and in agreeing to only a suitable degree of generalization. Another branch of topology, called *combinatorial topology*, considers, as we shall later explain in detail, a closed curve as a kind of curvilinear polygon and a surface as a kind of twisted polyhedron. It sets aside the fact that these surfaces and polygons are point sets; the surfaces are regarded as formulas determined by the sides and vertices of curvilinear polygons and by their agreed incidence relations. The topological study of surfaces is thus reduced to the combinatorial study of these formulas. The figures studied in combinatorial topology are *complexes*, which are generalizations of surfaces. Complexes constitute a class of figures sufficiently restricted for the application of combinatorial methods, but sufficiently large to embrace almost all interesting figures. Each complex lends itself to a combinatorial representation that allows topology to employ algebra. In effect, the study of the combinatorial formulas is related to linear algebra or the theory of groups. This gives the advantage of putting topology into closer relationship with the rest of mathematics.

Still, there is no rigid separation of the two disciplines of topology. For many topological questions, one adopts a compromise by using a mixed method. The combined use of set and algebraic methods often proves to be particularly profitable.[1]

Our book, as indicated by its title, is directed principally to combinatorial topology—accordingly, to the topology of complexes. Since complexes are generalizations of surfaces, we shall limit ourselves, in the other two chapters of this book, to the topology of surfaces.[2]

---

[1] For more details about the mixed method, see R. L. WILDER [28].

[2] In connection with the topology of surfaces, one can also consult: H. SEIFERT and W. THRELFALL [22], pages 130–148; F. LEVI [18], pages 40–90.

10. THE DEVELOPMENT OF TOPOLOGY

In the history of the development of the geometric sciences, topology is a relatively recent branch. It suffices to quote the following words, used by Gauss in 1833, to see the state of topology at that time: "Of the geometry of position, which Leibniz had initiated and to which it remained for only two geometers, Euler and Vandermonde, to throw a feeble glance, we know and possess, after a century and a half, very little more than nothing."

It was Riemann who, in searching for the deep relationships between the study of surfaces and the theory of functions, gave in 1851 the first application of topology to classical mathematics. With the work of Riemann, that of Möbius, Jordan, Schläfli, Dyck, Betti, and Kronecker came to form the first results of combinatorial topology. But these geometers constitute only the precursors of this science, which owes its most notable progress to H. Poincaré. The five papers which he published on this science furnished the point of departure for a great amount of research by various authors, among whom one can cite Brouwer, Lebesgue, Veblen, Alexander, Lefschetz, Alexandroff, and Hopf. With Poincaré commenced, in 1895, the systematic theory of combinatorial topology as we know it today. One can say that this illustrious geometer was the promoter of combinatorial topology.

On the other hand, independently of combinatorial topology, G. Cantor in 1879 founded set topology with his theory of sets. He defined the first fundamental topological notions in cartesian space of n dimensions and obtained the essential results on the topological structure of the line and the plane. Cantor's theory was soon employed and widely diffused by the French school of function theorists. To indicate how these ideas were diffused, first consider their possible application to sets, no longer of points, but of curves or of functions. This idea, due to Ascoli, Volterra, and Hadamard, appeared in 1884 and was directly related to the creation of functional calculus by Volterra in 1887. But later it was realized that knowledge of the nature of the elements of the set (points, curves, functions, etc.) is of little importance and that the essential thing is the topological structure among the elements of the set. Thus, in extricating that which is common to the topological properties of sets of points and sets of functions, one was led to generalize the concept of space and to introduce the topology of abstract spaces, spaces whose points are abstract elements of arbitrary nature. The first efforts in this direction were made by M. Fréchet in 1904. Since then, set topology has undergone a new development or become what is more properly described by the term *abstract topology* or *general topology*.[1]

A large part of recent topological research is devoted to the fusion of combinatorial topology and set topology. In the past few years this end has been attained to a high degree. All of Poincaré's theory is today generalized to very general sets. This progress has exercised a penetrating influence on all of contemporary topology.

---

[1] To get a general idea of this abstract topology, refer to F. Severi [23], pages 152–159. For a deeper study of this branch of topology, the reader can use M. Fréchet [7]; N. Bourbaki [4]; W. Sierpiński [24]; P. Alexandroff and H. Hopf [2], pages 23–124; G. T. Whyburn [27].

The extensive development of contemporary topology has made it actually impossible to conceive a theory of analysis that does not rest on a preliminary topological study. Topological facts, in spite of their apparent vagueness, are directly related to the most precise mathematical facts. In almost every branch of analysis or geometry, topological considerations often lead to the most fertile ideas. The introduction of topological methods in analysis goes back to Riemann, and their application was revived by Poincaré in his researches in differential equations and dynamical systems. Since then, the application of topological methods in the calculus of variations by Birkhoff, Morse, Lyusternik, and Schnirelmann, in the theory of differential equations by Birkhoff, Kellogg, and Schauder, in algebraic geometry by Lefschetz, Severi, and van der Waerden, and in differential geometry by many authors, constitutes a renovation of each of these mathematical disciplines [q].

CHAPTER TWO

Topological Notions About Surfaces

11. DESCARTES' THEOREM

Consider a polyhedron of elementary geometry. We shall designate the number of its vertices, edges, and faces by n_v, n_e, and n_f. Well known is the famous *Descartes' formula* (mentioned by Poincaré in the quotation on page vi), often attributed to Euler:[1]

$$(1) \qquad n_v - n_e + n_f = 2.$$

To determine the validity of this formula,[2] we first need a precise definition of polyhedron.

A *polyhedron* is a system of a finite number of polygons [r] (which are called *faces* of the polyhedron) which are situated in a mutual relationship such that the following four conditions are verified: (1) each pair of polygons of the system have no common interior points; (2) for each side of a polygon, there exist two and only two polygons having this side in common[3] (which is called an *edge* of the polyhedron); (3) each pair π, π' of polygons of the system can be joined by a sequence $\pi_1 = \pi, \pi_2,$ $\cdots, \pi_n = \pi'$ of polygons of the system in the sense that each of these has

---

[1] On the matter of priority to Descartes, see: W. KILLING and H. HOVESTADT [16], page 268; D. HILBERT and S. COHN-VOSSEN [11], page 254.

[2] The formula is usually written in the form $V + F = E + 2$, where V, E, F have, respectively, the same meaning as n_v, n_e, n_f in our notation.

[3] For example, if we remove one face of a cube, we no longer have a polyhedron.

a side in common with the following one;[1] (4) the polygons about any vertex can be placed in a cyclic order so that each consecutive pair have a common side passing through the vertex. According to this definition, when we speak of a polyhedron, we always mean the surface of the polyhedron.

Descartes' formula is not necessarily true of a polyhedron considered in this very general sense. The most important polyhedra are those called simple. A polyhedron is said to be *simple* if one can continuously deform it into the surface of a sphere. In Fig. 13 we have an example of a

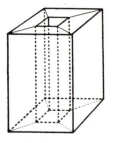

Figure 13

nonsimple polyhedron: it is the surface of a volume which one obtains by removing a rectangular parallelepiped from another rectangular parallelepiped of larger base but of the same altitude. This polyhedron contains a hole which every continuous deformation preserves, and which would cease to exist if the polyhedron should transform into a sphere.

Descartes' theorem can now be stated in a more precise form: *We have formula* (1) *for every simple polyhedron having n_v vertices, n_e edges, and n_f faces.* We are going to prove this theorem, using a highly intuitive approach pointed out by HILBERT and COHN-VOSSEN ([11], page 255).

Imagine a given simple polyhedron (that is to say, the surface of the polyhedron) as made of a sort of elastic material like India-rubber. We cut out an arbitrarily chosen face, remove it and stretch the other faces, without tearing or overlapping, into a planar piece in such a way that each face becomes a polygon of the same number of vertices as before

---

[1] For example, the figure formed by two tetrahedra with no common point (or having only a vertex in common) is not a polyhedron.

the operation and so that two faces having an edge or a vertex in common on the polyhedron become two polygons still having an edge or a vertex in common.[1] This is possible, since the elastic material of which the polyhedron is made allows distortion without tearing. (Of course, one cannot expect each polygon in the planar piece to be equal to the initial face of the polyhedron, but this is unimportant here.) We will obtain in this way a network of polygons in the plane [s]. (One sees in Fig. 14 the networks obtained by starting with the surfaces of a tetrahedron and a cube.)

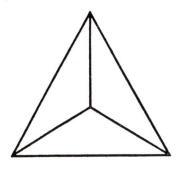 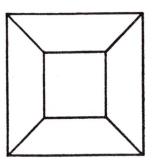

Figure 14

Now consider the network of polygons thus obtained in the plane. The number of vertices and the number of edges are the same as on the polyhedron. However, the number of polygons is one less than the number of faces of the polyhedron, since one of these faces has been cut out and removed. Therefore, the value of the expression $n_v - n_e + n_f$ for the network is one less than that of the same expression for the corresponding polyhedron. It suffices, then, to find the value of $n_v - n_e + n_f$ for the network.

If, in the network of polygons, there are some polygons which are not triangles, we divide these into triangles by means of diagonals (Fig. 15).

---

[1] It is important for the success of this process that the polyhedron be simple. For example, in operating on the nonsimple polyhedron of Fig. 13, after having removed a face, we cannot properly stretch the remaining faces into a planar piece, for there will have to be an overlapping of certain faces by others.

Each time that we add a diagonal, the number of polygons will be increased by one and the number of sides will be increased by one also, whereas that of the vertices will not be changed. Thus the addition of a diagonal has no effect on the value of the expression $n_v - n_e + n_f$. Therefore we can transform the network into another which is formed exclusively of triangles and which has the same value for $n_v - n_e + n_f$ as before.

Figure 15

On the other hand, we can obtain a similar network of triangles by starting with a single triangle of the network and using a finite number of operations of the following two kinds: one kind of operation consists of adding a new triangle with a side belonging to a triangle already obtained, by introducing the new vertex opposite this side and two new sides (Fig. 16); the other kind of operation consists of completing a triangle of which two sides have already been obtained, by introducing a new side (Fig. 17). But each of these two operations has no effect on the

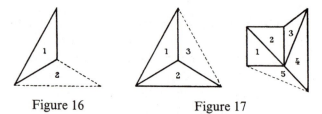

Figure 16   Figure 17

value of the expression $n_v - n_e + n_f$; an operation of the first kind, for example, increases by one the number of vertices as well as that of triangles, and increases by two the number of edges.

It follows that the value of the expression $n_v - n_e + n_f$ for the network of triangles is the same as for a single triangle. For the latter we have

$$n_v - n_e + n_f = 3 - 3 + 1 = 1.$$

Therefore, for the network of triangles and consequently for the network of polygons, we have

$$n_v - n_e + n_f = 1.$$

We thus have the relation (1) for all simple polyhedra [t].

12. AN APPLICATION OF DESCARTES' THEOREM[1]

As an application of Descartes' theorem, we are going to determine all the simple polyhedra that are regular in the sense of elementary geometry. Consider such a polyhedron. Suppose that there are h edges ending at each vertex and that each face has k sides. Again denote by n_v, n_e, and n_f the number of vertices, edges, and faces of the polyhedron. One easily establishes the two relations

$$(2) \qquad n_v h = 2n_e = n_f k.$$

By Descartes' theorem we then have

$$2n_e/h - n_e + 2n_e/k = 2,$$

or

$$(3) \qquad 1/n_e = 1/h + 1/k - 1/2.$$

On the other hand, we evidently have $h \geqq 3$ and $k \geqq 3$, no matter what polyhedron is considered. If the numbers h and k should both be greater than 3, we would have

$$1/n_e = 1/h + 1/k - 1/2 \leqq 1/4 + 1/4 - 1/2 = 0,$$

which is impossible. It follows that at least one of the numbers h and k is equal to 3. Suppose $k = 3$; we will have

$$1/n_e = 1/h - 1/6.$$

We then see that h can take on only the values 3, 4, and 5. For $k = 3$ and $h = 3, 4, 5$ we have, respectively, $n_e = 6, 12, 30$. Since equation (3) is

---

[1] One will find other consequences of Descartes' theorem in H. LEBESGUE [17].

symmetric in h and k, we also have $h = 3, k = 3, 4, 5, n_e = 6, 12, 30$. There are, then, for the simple regular polyhedra, at most five cases, which correspond to the following five systems of h, k, and n_e:

h	k	n_e
3	3	6
4	3	12
5	3	30
3	4	12
3	5	30

from which the relations (2) allow us to find n_v and n_f. But there actually are five kinds of simple regular polyhedra, namely the tetrahedron, the octahedron, the icosahedron, the cube, and the dodecahedron (Fig. 18),

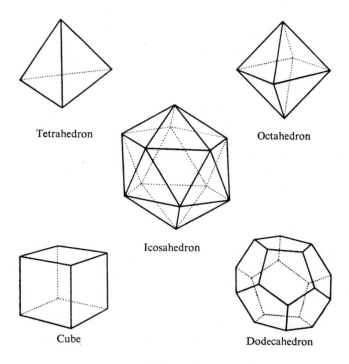

Tetrahedron

Octahedron

Icosahedron

Cube

Dodecahedron

Figure 18

which correspond, respectively, to the five preceding systems of h, k, and n_e:[1]

Names of the simple regular polyhedra	n_v	n_e	n_f	h	k
Tetrahedron	4	6	4	3	3
Octahedron	6	12	8	4	3
Icosahedron	12	30	20	5	3
Cube	8	12	6	3	4
Dodecahedron	20	30	12	3	5

Thus there exist five and only five kinds of simple regular polyhedra. It will be observed that they all are not only simple, but convex. These five kinds have been known ever since the famous Greek philosopher Plato, and it is for this reason that they are also called the *Platonic solids*.

Finally, it will be noticed that the above table is also what one arrives at when one ceases to suppose the polyhedra are regular—that is, if one seeks only the simple polyhedra for which the number of edges ending at a vertex is independent of this vertex and the number of edges of a face is independent of this face. Consequently, *the simple regular polyhedra can be determined by their topology, without bringing in considerations of size or measure* (in particular, it is not necessary to use the fact that their faces are regular polygons) [**u**].

13. CHARACTERISTIC OF A SURFACE

Let us now deform a simple polyhedron so as to transform it into the surface of a sphere; we obtain a division of the surface of the sphere into what can be called the curvilinear polygons corresponding to the faces of the polyhedron. Conversely, if one divides the surface of a sphere into a finite number of curvilinear polygons in such a way that for each curvilinear edge there exist two and only two curvilinear polygons having this edge in common, one can deform the sphere (like a ball of India-rubber) so as to transform it into a simple polyhedron whose

---

[1] Cf. D. HILBERT and S. COHN-VOSSEN [11], page 79. One will find in F. LEVI [18], Chap. VI, a very complete treatment of the regular polyhedra.

faces come from the curvilinear polygons drawn on the sphere. It follows that Descartes' theorem holds for all divisions of the surface of a sphere into curvilinear polygons, provided the division satisfies the condition stated above. In Fig. 19 is a division of the sphere into four curvilinear triangles. This division corresponds to the tetrahedron and is called the *tetrahedral division.*

Thus, for every division of the sphere into curvilinear polygons, we have $n_v - n_e + n_f = 2$, provided the division satisfies the above-stated condition. Consequently, the number 2 is assigned to the sphere in an intrinsic way, independently of the nature of the polygonal division; one says that the sphere has *characteristic* 2.

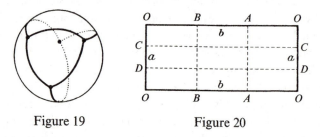

Figure 19   Figure 20

In place of the sphere, let us now take the torus. According to what we have said on page 4, each division of the torus into curvilinear polygons can be represented on a rectangle by considering the equivalent vertices (that is to say, those coming from the same vertices on the torus) and the equivalent edges. There is represented in this way, in Fig. 20, a division of the torus into nine curvilinear rectangles. For this division we have

$$n_v = 9, \quad n_e = 18, \quad n_f = 9,$$

and consequently

$$n_v - n_e + n_f = 0;$$

therefore Descartes' formula does not hold for the torus. But it can be shown that, when the torus is divided into a finite number of curvilinear polygons in such a way that for each (curvilinear) edge there are two and only two curvilinear polygons having this edge in common, we always have

$$n_v - n_e + n_f = 0.$$

The number 0 is thus assigned to the torus independently of the way in which the torus is divided. We say, then, that the characteristic of the torus is 0.

More generally, we can define the characteristic for any surface (see page 65), and we shall see (page 68) that the characteristic of a surface is a topological invariant—that is, two homeomorphic surfaces always have the same characteristic. One notices, for example, that the torus is homeomorphic to the polyhedron of Fig. 13. For that polyhedron we have

$$n_v = 16, \quad n_e = 32, \quad n_f = 16,$$

whence

$$n_v - n_e + n_f = 0.$$

14. UNILATERAL SURFACES

When an ant is put on the surface of a sphere, it cannot pass, without piercing the surface, from the exterior of the sphere to the interior of the sphere by following a path on the surface. Similarly, if an ant is placed on the surface of a hemisphere, it cannot pass, without crossing over the edge of the hemisphere, from one side of the surface to the other by traveling along a path on the surface. A surface having this property is said to be *bilateral*.

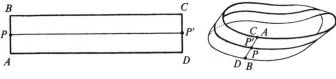

Figure 21

There exist surfaces which do not possess this property. The Möbius band furnishes us the simplest example: take a rectangular piece of paper $ABCD$ (Fig. 21) whose base AD is quite long compared with its altitude AB. After twisting this strip through an angle of 180°, rejoin the sides AB and CD so that points A and C coincide, as well as points B and D (Fig. 21). The surface so obtained is called a *Möbius band*[1] [v]. One

---

[1] Actually, it was J. B. Listing who first studied this surface in 1861, four years before Möbius. Concerning this question of priority, see TIETZE [26], page 155, note.

sees that the edge of the band is composed of a single closed curve. An ant can pass, without crossing over the edge of the band, from one side of this surface to the other by following a curve on the band. Consider, for example, the median line PP' of the rectangle; it becomes the closed curve PP' on the band. P and P' are two superposed points, one on one side of the band and the other on the other side of the band (Fig. 21). Thus the Möbius band does not, like the hemisphere, have two sides which are separated by the edge and of which one can be colored red and the other blue in such a way that the two colors touch only along the edge. We say that the Möbius band, in the space of elementary geometry, is a *unilateral surface*.

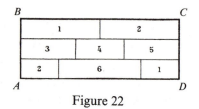

Figure 22

Let us point out in passing that the chromatic number (page 3) as well as the maximum number of pairwise adjacent regions (page 5) have been completely determined for the Möbius band. H. TIETZE [26] has proved that the chromatic number of the Möbius band is 6. It follows, then, that the maximum number of pairwise adjacent regions is at most 6. It is actually 6. In fact, Tietze pointed out that there exist, on the Möbius band, six regions touching each other along a curve. To show this last fact, we first divide a rectangle $ABCD$ (Fig. 22) into three horizontal rectangular strips; then we divide the first strip into two regions 1 and 2; the second strip into three regions 3, 4, 5, region 4 being adjacent to regions 1 and 2; the third strip into regions 2, 6, 1, region 6 being adjacent to regions 3, 4, 5. Then, when (after twisting the rectangle through an angle of 180°) we join the vertical sides AB and CD to obtain the Möbius band, the two regions numbered 1 (or 2) on the rectangle form only one on the band. We thus have six regions on the band; they are pairwise adjacent. Thus the maximum number of pairwise adjacent regions for the Möbius band is 6. By what we said on page 6, 6 is also the maximum number of points on the band that can be pairwise joined by mutually nonintersecting curves drawn on the band [w].

The Möbius band is not a closed surface, since it has an edge. One may wonder if there exists any closed unilateral surface in the space of elementary geometry. Such a surface, having no edge, has neither an interior nor an exterior. The surface, then, must necessarily penetrate itself in certain places.[1] Here is an example.

Imagine that the surface of a sphere is made of India-rubber. First mark a small quadrilateral *ABCD* on the surface and cut it out (Fig. 23). Next deform the remaining surface so that it takes the form of Fig. 24. Close the opening by attaching *AB* to *CD* and *DA* to *BC*. We thus obtain the surface of Fig. 25. This is a closed surface having a curve of

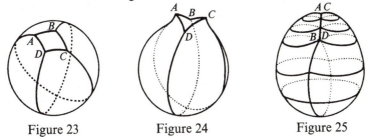

| Figure 23 | Figure 24 | Figure 25 |

penetration *AB*. (Except for the two points *A* and *B*, each point on this curve of penetration must be considered as a pair of two distinct points.) And it is precisely due to this curve of penetration that the surface is unilateral.[2]

15. ORIENTABILITY AND NONORIENTABILITY

Unilateral surfaces in the space of elementary geometry can be characterized by another topological property. Imagine that we have marked a small circumference or small ring around each point of the surface (except those on edges, if such exist). Now try to give each of these circumferences a definite sense of cyclic direction, so that two circumferences with neighboring centers receive the same sense. If this is possible, we shall say that the surface is *orientable*; otherwise we shall say that it is *nonorientable*.

---

[1] Of course, this purely intuitive observation is not a proof.

[2] Figs. 23, 24, 25 are taken from HILBERT and COHN-VOSSEN [11], page 277.

A unilateral surface cannot be orientable. In fact, on each unilateral surface there exists at least one closed curve (not crossing any edge, if such exist), along which an ant can pass from a point situated on one side of the surface to a superposed point on the other side—for example, the median curve PP' on the Möbius band (Fig. 21). If a moving point travels along this curve from P to P' and if circumferences around neighboring positions of the moving point are oriented in the same sense, the sense of the circumference around P' is necessarily opposite to that of the circumference around P (Fig. 26). Now P and P', being

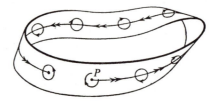

Figure 26

two superposed points, are the same point. The Möbius band is thus nonorientable. This pecularity holds only for unilateral surfaces.

If the Möbius band is cut along the closed curve PP', it will not separate into two pieces, as the reader (who is usually surprised) will see by making the experiment [x]. The closed curve PP' possesses the following remarkable property: if a moving point travels on the band close to and throughout the length of PP', it will pass from one side of the curve to the other without crossing it, so that if the surface of a continent assumes the form of a Möbius band and is traversed by a river following the curve PP', one can walk from one bank of the river to the opposite bank without crossing over the river; this can be expressed by saying that the river has only one bank [y]. We can then also say that the closed curve PP' on the Möbius band has only *one bank*. It is easy to see that the existence of at least one closed curve having only one bank is a necessary and sufficient condition for the surface to be unilateral. Using this fact, it can be shown (see DE KERÉKJÁRTÓ [14], pages 137–138) that every bilateral surface is orientable.

Now let us divide a surface into curvilinear polygons (which we can also call faces) in such a way that each polygonal side is common to at most two faces. (If the surface is closed, each side is common to exactly two faces; if the surface is not closed, each side on an edge of the surface belongs to only one face and each of the other sides is common to two faces.) By means of such a polygonal division, orientability can be defined in another way.

Consider, first, the surface of a tetrahedron $ABCD$ or that of a sphere having a tetrahedral division. The surface, being bilateral, is orientable; that is, by drawing a small circumference about each of its points, we can give a definite cyclic sense to each of these circumferences so that

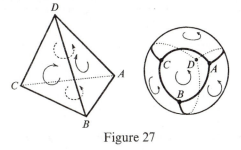

Figure 27

any two circumferences with very close centers have the same sense. Having thus determined the cyclic sense of each little circumference, concentrate, now, on a face of the tetrahedron, say face BCD. One sees that, for any two points of this face, the two circumferences about these points necessarily have the same sense. Consequently, instead of marking all the little circumferences (about the points of the face BCD) with the same cyclic sense, we can mark only one circumference, or curved arrow, in the interior of triangle BCD, the arrow indicating the cyclic sense that is common to all the little circumferences. This can further be indicated by a cyclic order of the vertices of the triangle, say by the order BCD. (This means that all the little circumferences around the points of triangle BCD have the same sense as that in which a point turns on the perimeter of the triangle by starting from B and passing successively through C and D to return to the point of departure B.) This process can naturally also be applied to the other faces of the tetrahedron. But, bearing in mind the hypothesis that two little circumferences with very close centers have the same cyclic sense, we see that, if the face BCD receives the order BCD, the other faces necessarily have the orders CAD, ABD, BAC (Fig. 27). Consequently, each polygonal side common to two faces will then receive opposite senses, one for one face and the other for the other face. This is what is called *Möbius's rule of edges*.

More generally, a surface having a polygonal division (satisfying the above condition)[1] is said to be orientable if one can fix a sense of direction on each face so that each side common to two faces receives two opposite senses of direction, one for one face and the other for the other face. We see that this definition is equivalent to the preceding one and that, consequently, it is independent of the polygonal division employed. Thus the property of being orientable or not is an intrinsic

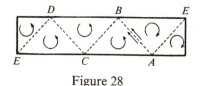

Figure 28

property of the surface; it can be defined with the aid of a polygonal division, but is basically independent of this. Furthermore, we shall see (page 68) that this property is a topological invariant—two homeomorphic surfaces are both orientable or both nonorientable.

As an example, again take the Möbius band and divide it into five triangles *ABC, BCD, CDE, DEA, EAB* (Fig. 28). If, for instance, we give to triangle *ABC* the sense of direction *ABC*, we necessarily have for triangle *BCD* (if we apply Möbius's rule of edges) the sense *BCD*, and consequently we will similarly have the senses *CDE, EDA, EAB* for the other triangles, considered successively in this order. It results that the edge *AB* is twice oriented in the same sense. Thus the Möbius band is nonorientable.

As another example, take an octahedron *ABCDEF* (Fig. 29) and consider the figure *P* formed by the four faces *AED, EBC, CFD, ABF* and the three squares *ABCD, EBFD, AECF*. We thus have seven polygons, four of which are triangles and three are squares. Their sides and their vertices coincide with the edges and vertices of the octahedron. The diagonals *AC, BD, EF* are not considered as sides, but only as lines of

---

[1] That is, each polygonal side is common to one or two faces according as it is or is not on an edge of the surface.

intersection where two polygons penetrate one another. Each edge of P is common to two and only two polygons. Except for the vertices, each point on a line of penetration should be regarded as a pair of two distinct points. It follows that the seven polygons are considered as without any common interior points. It is easily verified that the seven polygons form a polyhedron. This polyhedron is called the *heptahedron of C. Reinhardt.*[1] It is nonorientable. In fact, consider, for example, the four

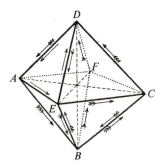

Figure 29

faces AED, $EBFD$, EBC, $ABCD$, each consecutive pair of which have a common edge. If we fix, let us say, the sense of direction of face AED in the order AED, the senses of direction of the other three faces are successively determined by Möbius's rule of edges; these three faces will then necessarily have the senses of direction given by the orders $DEBF$, BEC, $BCDA$. Then the edge AD is twice oriented in the same sense (Fig. 29).

16. TOPOLOGICAL POLYGONS

Since every polygon is homeomorphic to a circular disk (that is, to the surface of a circle), we will now say that a circular disk becomes a

---

[1] The heptahedron of Reinhardt cannot be a simple polyhedron, for Descartes' formula does not hold—we have, in fact, $n_v - n_e + n_f = 6 - 12 + 7 = 1$. We cannot, then, continuously deform the heptahedron into a sphere. On the other hand, it is interesting to note that the heptahedron can be continuously deformed into a surface whose equation, related to a rectangular coordinate system, is

$$y^2z^2 + z^2x^2 + x^2y^2 + xyz = 0.$$

This surface is called *Steiner's calyx surface* or *Steiner's Roman surface.*

topological polygon, or a *curvilinear polygon*, or more briefly a *polygon*, if its circumference is divided into a certain number r (≥ 2) of arcs, called *sides*, by means of the same number r of points, called *vertices*. Thus the polygon is determined by the circular disk and these r arcs. Any figure homeomorphic to a topological polygon will bear the same name.[1] When $r > 2$, a topological polygon is always homeomorphic to a convex rectilinear polygon of the kind studied in elementary geometry. There exists no rectilinear polygon of two sides, but Fig. 30 represents a curvilinear polygon of two sides. According to the definition itself, rectilinear polygons are only particular cases of topological polygons. Henceforth, whenever we speak of a polygon without indicating whether it is rectilinear or not, it will be understood that we mean a topological polygon.

17. CONSTRUCTION OF CLOSED ORIENTABLE SURFACES FROM POLYGONS BY IDENTIFYING THEIR SIDES

We are going to construct certain closed surfaces by deforming polygons and pasting—that is, making certain sides coincide. These operations clearly are not homeomorphic, since two distinct points can coincide after the pasting. But the construction will illustrate the possibility of decomposing certain surfaces into one or more polygons satisfying certain conditions. And it is this possibility that will suggest a rule allowing us to restrict the subject of topology of surfaces (see further, page 49).

If we cut a spherical surface, made of a flexible material, along an arc whose extremities are P and Q, and then separate the two lips of the cut, we obtain an open surface which is a curvilinear polygon of two sides, for, if we wish, we can flatten it so as to transform it into the surface represented in Fig. 30. Conversely, to reform the sphere starting with this surface, we have only to close the two sides of the polygon, as when one snaps together the two sides of a purse about its hinges P and Q. In other words, the sphere can be obtained from a polygon of two sides (Fig. 30) by deforming the surface which the polygon bounds so as to

---

[1] This definition conforms to the general principle of considering any two homeomorphic sets as topologically equivalent.

make coincide—we more briefly say to identify—the two sides in such a way that the heads of the arrows drawn on them in Fig. 30 coincide. If we give a definite sense of direction to the perimeter of the polygon [which can be indicated by a new arrow placed in the interior of the polygon (Fig. 30)], we can assign to each side a + or a − sign according as the arrow on that side is or is not in the same sense as that of the

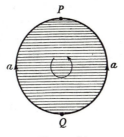

Figure 30

perimeter. We can then symbolically represent the polygon of two sides by

$$(4) \qquad\qquad a^+ a^-.$$

Similarly, we have seen (page 4) that the torus can be obtained from a polygon of four sides (Fig. 4) by deforming the polygon so as to make the pairs of opposite sides coincide.

In Fig. 4, the pairs of opposite sides of the rectangle are designated by the same letter. This indicates that the pairs of opposite sides are equivalent—that is, are to be pasted together. But there are two ways of pasting a pair of sides. For example, when one starts by pasting the two sides a, one can make them coincide so as to obtain a cylinder or so as to obtain a Möbius band. To make the chosen mode of pasting precise, we have marked (Fig. 4) an arrow on each side so that after the pasting the heads of the arrows on two equivalent sides coincide. On the other hand, we have marked, in the interior of the rectangle, a curved arrow which indicates the arbitrarily chosen cyclic sense of direction of the perimeter of the rectangle. If we now write in a line the letters representing the sides of the rectangle, in the cyclic order determined by the curved arrow, and if we attach to each letter a + or a − sign according as the arrow on the corresponding side is or is not in the same sense as that of the perimeter, we will then have a symbolic representation of the rectangle and the equivalence of its sides:

(5) $$a^+b^+a^-b^-.$$

Now let us start with the polygon of eight sides shown in Fig. 31, in which pairs of equivalent sides are designated by the same letter. In this Fig. 31, as in Figs. 30 and 4, we will again utilize two kinds of arrows. The arrows on the sides are still employed to indicate that the placing of two equivalent sides in coincidence should be done so that the heads of the arrows on these sides coincide; this first kind of arrow, suffices, then, to describe the mode of identification of the sides. The curved arrow situated in the interior of the polygon is introduced only to obtain a symbolic representation of the figure. Having marked a curved arrow, we first write in a line the letters representing the sides of the polygon, in the cyclic order determined by this curved arrow. Then, to indicate the mode of identification of the equivalent sides, we attach to each letter the sign + or − according as the arrow on the corresponding side is or is not in the same sense as that of the perimeter (the sense of the perimeter being given by the curved arrow). The polygon of Fig. 31 can thus be symbolically represented[1] by

(6) $$a^+b^+a^-b^-c^+d^+c^-d^-.$$

Figs. 31 through 36, taken from HILBERT and COHN-VOSSEN [11], page

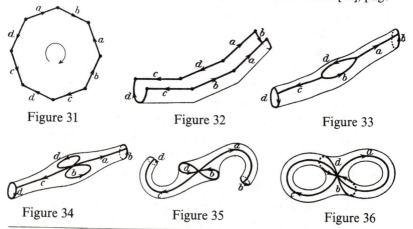

Figure 31   Figure 32   Figure 33

Figure 34   Figure 35   Figure 36

---

[1] We shall return on pages 51–53 with more details on the symbolic representation and on the employment of arrows.

265, successively show how, starting with this polygon, one forms the surface called the *generalized torus with two holes*.

Generalizing the symbolic representations (5) and (6), we now consider the symbolic representation

(7) $a^+b^+a^-b^-c^+d^+c^-d^-e^+f^+e^-f^-.$

It represents a polygon of twelve sides. Figs. 37 through 40, also taken from D. HILBERT and S. COHN-VOSSEN [11], page 265, successively show how one transforms it into a *generalized torus with three holes*.

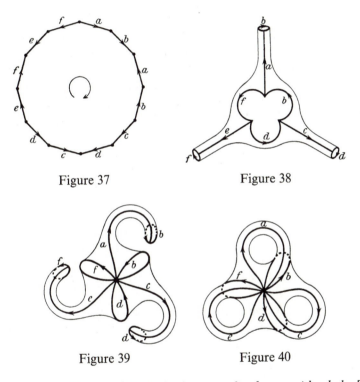

Figure 37   Figure 38

Figure 39   Figure 40

More generally, one will obtain the *generalized torus with p holes* [z] by starting with the polygon of $4p$ sides

(8) $a_1^+b_1^+a_1^-b_1^-a_2^+b_2^+a_2^-b_2^- \cdots a_p^+b_p^+a_p^-b_p^-.$

By starting with polygons, we have thus constructed, by deformation and pasting, the sphere, the torus, and the generalized torus with a finite number of holes. But this construction is of much more general application; it is clear that it applies to all surfaces obtained by continuously deforming any of the preceding—that is, the construction applies to surfaces of extremely varied form from the metrical point of view. On the other hand, we will see later that all these surfaces are orientable.

18. CONSTRUCTION OF CLOSED NONORIENTABLE SURFACES FROM POLYGONS BY IDENTIFYING THEIR SIDES

Let us now return to the surface in Fig. 25. We have seen that it can be obtained, by deforming and pasting, from a sphere from which a small quadrilateral has been removed. When a small quadrilateral is removed from a sphere, the resulting surface is a polygon of four sides (Fig. 23). We see, then, that the surface of the figure is obtained from the four-sided polygon $a^+b^+a^+b^+$ (Fig. 41). By combining the two sides a and b

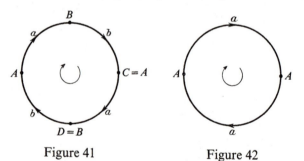

Figure 41   Figure 42

into one side, which we designate by just the letter a (Fig. 42), we see that the surface in Fig. 25 can be obtained from a polygon of two sides. The symbolic representation of this polygon is

$$(9) \qquad\qquad a^+a^+.$$

From Fig. 42, we can say that the surface in Fig. 25 is obtained from a circular disk by identifying the diametrically opposite points of its circumference.

Now consider the four-sided polygon (Fig. 43):

$$(10) \qquad\qquad a^+a^+b^+b^+.$$

Let us cut this along the diagonal c into two triangles (Fig. 43) and then attach the triangles along the two sides b. We then obtain the four-sided polygon $a^+c^+a^-c^+$ (Fig. 44). By attaching the two sides a, we obtain a

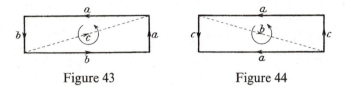

Figure 43   Figure 44

surface of the form of a tube of which the two ends are c (Fig. 45). But to identify the two ends, it is necessary to match the arrows on them. (In fact, the senses of direction of the two ends c are opposite, which constitutes the only distinction between the case under consideration and the case of the torus.) To accomplish this, we can proceed as follows. Deform the tube so that one of the two ends is a little smaller than the other and penetrate this smaller end through the wall of the tube. Then

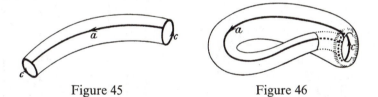

Figure 45   Figure 46

bend the larger end a little toward the interior and the smaller end a little toward the exterior, permitting us to attach them according to the required mode of identification (Fig. 46). The surface so obtained is called the *Klein surface* or the *nonorientable torus* [aa]. It is, indeed, nonorientable, for it is easy to see that it is unilateral.

In forming closed nonorientable surfaces, it is convenient to utilize an open surface called a *cross cap* (Fig. 49), which is nothing but the top half of the surface in Fig. 25. The cross cap is obtained by starting with a circular ring and identifying the diametrically opposite points of one of its edges, as is easily verified in Figs. 47 through 49. (The two arcs a of Fig. 47 are respectively deformed into the arcs ADC and CBA of Fig. 48.)

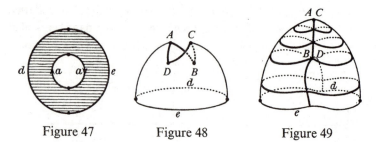

Figure 47   Figure 48   Figure 49

On the other hand, by identifying the diametrically opposite points of one of the two edges of a circular ring, one also obtains a Möbius band. First cut the circular ring (Fig. 50) into two parts (Fig. 51), deform these two parts into two rectangles, and then paste the two rectangles along the equivalent sides a (Fig. 52), thus obtaining a single rectangle (Fig. 53, where the side f is merely the union of sides b and c of Fig. 52). After twisting this rectangle, attach the two sides f to form a Möbius band. It

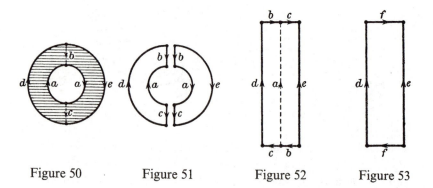

Figure 50   Figure 51   Figure 52   Figure 53

is seen that one of the two edges of the circular ring (more exactly, that one whose diametrically opposite points are identified) has become the median curve of the Möbius band and that the other edge has become the edge of the band.[1]

---

[1] Conversely, as the reader will be able to see by experiment, if the Möbius band is cut along its median curve, it will not separate into pieces but will become a surface which essentially has the metrical form of a sort of circular ring situated in space in a doubly twisted fashion.

From the relationship subsisting between the cross cap and the circular ring on the one hand, and between the Möbius band and the circular ring on the other hand, one immediately concludes that there exists a homeomorphism between the cross cap and the Möbius band such that the edge of the one corresponds to the edge of the other and the curve of penetration on the cross cap corresponds to the median curve on the Möbius band. The fact that the cross cap has a line of penetration while the Möbius band does not depends only on the situation of the surfaces in space.

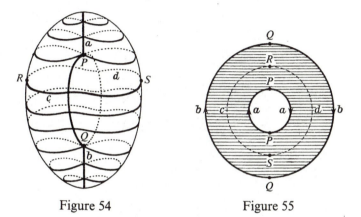

Figure 54   Figure 55

If a circular hole is made in the surface of a sphere and then closed by pasting a cross cap over it edge to edge, the surface in Fig. 25 is regained. This is why that surface is called a *sphere with a cross cap*.

Now make two circular holes in the surface of a sphere and close each of them by means of a cross cap. The surface so obtained is called a *sphere with two cross caps*. It is clear that this surface can also be directly obtained by pasting two cross caps edge to edge (Fig. 54).

Next let us take a circular ring whose interior circumference is formed of two arcs denoted by a and whose exterior circumference is formed of two arcs denoted by b (Fig. 55). By cutting the circular ring along a concentric circumference (formed of two arcs denoted by c and d), we obtain two circular rings. In the smaller of these two new circular rings, we identify the diametrically opposite points of the interior circumference, which gives us a cross cap with edge formed by the two arcs c

and d. In the larger of the two circular rings, we identify the diametrically opposite points of the exterior circumference, which gives us another cross cap with edge again formed by the two arcs c and d. Finally, if we attach these two cross caps, edge to edge, we will have a sphere with two cross caps. Thus a sphere with two cross caps is obtained from a circular ring by identifying the diametrically opposite points of each of its circumferences.

More generally, if we make q circular holes in the surface of a sphere and close each of them by pasting on a cross cap, the resulting closed

Figure 56

surface is called a *sphere with q cross caps*. This surface possesses q curves of penetration; it is therefore nonorientable. If from a circular disk we remove $q - 1$ small circular disks, we obtain a region bounded by q circumferences. It is easy to see that the identification of diametrically opposite points of each of these q circumferences gives us a sphere with q cross caps.

The region bounded by q circumferences can clearly be considered as a union of polygons. It follows, then, that a sphere with q cross caps can be obtained by starting with some polygons and identifying their sides.

Finally, observe that we can decompose the Klein surface (Fig. 46) into two Möbius bands (Fig. 56). (Each of these bands is situated in space in such a way that it penetrates itself.) Thus the Klein surface can be obtained by pasting two Möbius bands edge to edge [bb]. But we have seen that a Möbius band is homeomorphic to a cross cap. It then follows that the Klein surface is homeomorphic to a sphere with two cross caps.

19. TOPOLOGICAL DEFINITION OF A CLOSED SURFACE

In the preceding we have spoken of surfaces, but we have not yet explained exactly what we mean in topology by the word *surface*. We are now going to give a precise topological definition of a surface. For simplicity, we shall limit ourselves to so-called *closed*[1] surfaces.

In the last two sections, we have constructed some closed surfaces from polygons by identifying their sides. This shows us the possibility, by operating in the reverse order, of decomposing any one of these surfaces into one or more polygons satisfying certain conditions. These surfaces are so varied from the point of view of their metrical form that their very generality allows us in the following to limit the consideration of surfaces to those that can be cut up into polygons subjected to certain conditions that we are going to make precise. It is quite natural to choose these conditions so as to generalize the conditions imposed on the polyhedra defined on page 21. Because of this limitation, it would be natural to call these surfaces *closed polyhedral surfaces*. But, since it can be shown that this limitation is more apparent than real, we shall retain only the term *closed surfaces*.

We have given on page 21 a definition of the word *polyhedron* that gives us a more general interpretation than is meant by this word in elementary geometry. From now on, the term *topological polyhedron*, or more briefly *polyhedron*, will be interpreted in the following even wider sense: *A topological polyhedron (or a polyhedron) is a system of a finite number of topological polygons, the system of these polygons being subjected to the following four conditions:* (1) *any two polygons of the system have no common interior point;* (2) *the sides of the polygons of the system coincide in pairs* (from which it follows that the total number of sides of these polygons is even); (3) *the polygons of the system cannot be separated into two partial disjoint*[2] *systems such that every side of a polygon of each partial system coincides with a side of a polygon of the same partial system;*[3] (4) *for any vertex of a polygon of the system, the polygons of the system having this vertex in common can be arranged in a*

---

[1] For a definition of an *open* surface, consult DE KERÉKJÁRTÓ [14], page 164.

[2] That is, two partial systems with no common polygon.

[3] The third condition implies that the polyhedron is of one piece.

cyclic order $\pi_1, \pi_2, \cdots, \pi_n$ *such that* π_i *and* π_{i+1} ($1 \leqq i \leqq n$, *where* π_{n+1} $= \pi_1$) *have a common side passing through the considered vertex.*

The polygons constituting a polyhedron are called the *faces* of the polyhedron. Two superposed sides form an *edge* of the polyhedron. A point of the polyhedron is called a *vertex* if it is a vertex of one of the polygons constituting the polyhedron.

Notice right off in what way the present definition of a polyhedron generalizes that given on page 21. First of all, in the definition on page 21, the polygons are planar, whereas in the present definition the polygons are considered as topological polygons. Furthermore, in the present definition the system of polygons defining the polyhedron can consist of only one polygon (that is, a polyhedron can have only one face), and two coinciding sides can belong to the same polygon, which is not the case in the definition on page 21.

We define a *closed surface* as the set of points belonging to a polyhedron.

Like the notion of (topological) polygon, that of (topological) polyhedron is invariant under all homeomorphisms. It thus follows that any set homeomorphic to a closed surface is again a closed surface, whence the property of a set of being a closed surface is a topological invariant.

One can divide the same closed surface in several ways into polygons subject to the preceding four precise conditions. Such a decomposition of a closed surface is called a *polygonal division*. Consequently, the same closed surface can give rise to several polyhedra. For example, we can consider the sphere as a curvilinear cube, or a curvilinear tetrahedron, or a curvilinear octahedron, etc.; we thus have different polyhedra. We must, then, distinguish between the notion of a closed surface and that of a polyhedron. A closed surface is a set of points which yields to a polygonal division—what is important for the closed surface is, then, the possibility of a polygonal division and not the division itself; a polyhedron is a closed surface for which the polygonal division is given. Whereas a closed surface is a set of points, a polyhedron is moreover a set of figures, more precisely a set of polygons which are situated in a certain mutual relationship.

Now, without doubt, what interests us is the notion of a closed surface and not that of a polyhedron. The latter is not even a secondary notion; but it allows us to consider a closed surface as a system of polygons, which is very convenient—as we shall see in Chapter Three—for the topological study of closed surfaces. Moreover, the general method (*combinatorial method*) of combinatorial topology stems from this mode of generation in defining surfaces as well as the more general sets which the subject studies.

CHAPTER THREE

Topological Classification of Closed Surfaces

20. THE PRINCIPAL PROBLEM IN THE TOPOLOGY OF SURFACES

In Chapter Two we pointed out some important topological notions concerning surfaces; those notions form the base on which rests the theory that we propose to develop here.

The principal problem in the topology of closed surfaces is the search for the topological invariants of each closed surface so that we can tell if two arbitrarily given closed surfaces are or are not homeomorphic.

There is an enormous variety of surfaces homeomorphic to a given surface. The topological point of view permits us to simplify the above problem by arranging the closed surfaces into pairwise disjoint classes, so that two homeomorphic surfaces belong to the same class and two nonhomeomorphic surfaces belong to different classes. Thus the principal problem is to enumerate all the distinct classes of closed surfaces and to characterize each class by topological invariants. We shall see in this chapter that the characteristic and the property of being orientable or not are two topological invariants that suffice to characterize each class. In other words, two closed surfaces are homeomorphic if and only if they have the same characteristic and they are both orientable or both nonorientable. This will completely resolve the principal problem in the topology of closed surfaces.

21. PLANAR POLYGONAL SCHEMA AND SYMBOLIC REPRESENTATION OF A POLYHEDRON

By its definition, a polyhedron is a figure in space formed by a finite number of polygons situated in a certain mutual relationship. So what is important for a polyhedron are its faces and their mutual relationship. These can be represented, using certain notations that we are going to make precise, by a system of polygons in a plane. This system of planar polygons will constitute what we shall call the *planar polygonal schema* of the polyhedron. This graphic schema, in turn, lends itself to a symbolic representation which allows us to calculate algebraically and consequently to facilitate the study of polyhedra.

Given a polyhedron, let us associate a polygon in the plane with each of its faces, this polygon being subject to the single condition of having the same number of sides as the associated face of the polyhedron. (The metrical form of this planar polygon is of no importance here.) We thus obtain a finite number of polygons in the plane, each of which is separated from the others. We are going to couple the sides of these planar polygons in pairs[1] in such a way that two coupled sides in the plane lead to the same edge of the polyhedron; two such sides in the plane will be called *equivalent* or *identified*.[2] In an analogous fashion, two vertices of the planar polygons will be called equivalent or identified if they lead to the same vertex of the polyhedron. But, while there are exactly two equivalent sides in the plane corresponding to each edge of the polyhedron, the number of equivalent vertices in the plane corresponding to a vertex of the polyhedron may not be two.

Now denote each couple of equivalent sides in the plane by the same letter. This does not yet suffice to describe the mutual relationship of the faces on the polyhedron—for example, if two sides AB and CD in the plane are denoted by the same letter, we know only that they lead to a common edge of the polyhedron, but we do not know if the vertex of the polyhedron corresponding to A coincides with the vertex of the poly-

---

[1] "To couple the sides of the polygons in pairs" here means "to make the sides of the polygons correspond in pairs," still leaving the planar polygons separated from one another.

[2] Two equivalent sides in the plane may or may not belong to the same polygon.

hedron that corresponds to C or to D. (This is seen, for example, if we compare the torus obtained from the rectangle of Fig. 4 with the Klein surface obtained from the rectangle of Fig. 44.) To make this precise, which is important in order to describe the mutual relationship of the faces of the polyhedron, we shall proceed as follows: each side in the plane will be oriented by placing an arrow on it in an arbitrary sense, except that for all couples of equivalent sides the heads of the two arrows on them coincide when the two sides are identified in forming the corresponding edge of the polyhedron. With these indications concerning the equivalence of sides, the set of polygons form a *planar polygonal schema* of the polyhedron. The curved arrows, previously mentioned on page 38, will be subsequently introduced in order to give a unique symbolic representation. Thus in the following we mark the curved arrows in a planar polygonal schema only if we have to deal at the same time with the symbolic representation of the schema.

The planar polygonal schema is determined, from the combinatorial view, by the polyhedron. In other words, the number of polygons and the number of sides of each polygon, along with the equivalence relations between the sides, are determined by the polyhedron. Of course, we can change the sense of the arrows, provided we change together those on two equivalent sides.

We are now going to introduce a symbolic representation. Arbitrarily let us give a cyclic sense of direction to the perimeter of each polygon in the planar polygonal schema by means of a curved arrow placed in the interior of the polygon. We can then attach a $+$ or a $-$ sign to each side according as the arrow on that side is or is not in the same sense as the perimeter. The sense of direction of the perimeter of a polygon determines at the same time a cyclic order of its sides. We will then obtain a *symbolic representation* of the polyhedron by writing for each polygon of the schema, in a line and in their cyclic order, the letters representing the sides of the polygon and by attaching a $+$ or a $-$ sign to each of these letters according to the above-mentioned convention.

As an example, consider the polyhedron obtained from the sphere by a tetrahedral division (Fig. 27). It gives rise to a planar polygonal schema formed of the four triangles of Fig. 57, along with their equivalence relations fixed by means of the arrows marked on the sides. By marking a curved arrow in the interior of each triangle (Fig. 57), we have the following symbolic representation:

$$\begin{cases} a^+b^+c^+, \\ c^-e^-d^-, \\ a^-d^+f^-, \\ f^+e^+b^-. \end{cases}$$

In a symbolic representation of a polyhedron, each line gives the cyclic order of the sides of a face of the polyhedron, and the designation of the sides and the $+$ and $-$ signs indicate the mutual relationship of the faces. Consequently, if two polyhedra P_1 and P_2 have the same

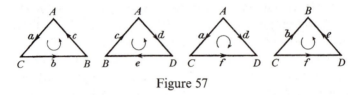

Figure 57

symbolic representation, they are composed of the same number of faces having, pairwise, the same symbolic representations. These faces are then pairwise homeomorphic. Furthermore, one can even establish a biunique correspondence ϕ between P_1 and P_2 so that, in the correspondence ϕ, each face of P_1 corresponds precisely to the face of P_2 having the same symbolic representation, and so that the correspondence established by ϕ between each pair of such faces is not only biunique, but also bicontinuous. In this biunique correspondence ϕ, each edge common to two faces of P_1 clearly corresponds to the edge common to the corresponding faces of P_2. We see, then, that ϕ is not only bicontinuous between each pair of faces of P_1 and P_2, but also bicontinuous between P_1 and P_2. Therefore, if two polyhedra have the same symbolic representation or the same planar polygonal schema, the two closed surfaces determined by these polyhedra[1] are necessarily homeomorphic.

The same polyhedron can give rise to different forms of symbolic

---

[1] A closed surface determined by a polyhedron is the set of points belonging to this polyhedron.

representation. In fact, we can first of all permute the letters in any line in a cyclic fashion. (For example, in the above illustration, we can write the first line in the form $b^+c^+a^+$ or $c^+a^+b^+$.) Furthermore, if we change the sense of the arrows on two equivalent sides in the planar polygonal schema, we change the two signs attached to the letter representing these two sides in the symbolic representation. (In the above illustration, if we otherwise orient the two sides b, the first and fourth lines in the symbolic representation become, respectively, $a^+b^-c^+$ and $f^+e^+b^+$.) And if we change the cyclic sense of the perimeter of a polygon in the planar polygonal schema, the symbolic representation will undergo the following change: the letters in the line representing this polygon will be arranged in the reverse cyclic order and the signs attached to the letters in this line will all be changed. (If we change the cyclic sense of the first triangle of Fig. 57, the first line in the symbolic representation becomes $c^-b^-a^-$.)

22. ELEMENTARY OPERATIONS

Given two polyhedra, what are the conditions that must be satisfied in order that the two closed surfaces determined by them be homeomorphic? With a view toward solving this problem, one is led to think of the following question: Given two polyhedra P_1 and P_2, if we know that the two closed surfaces determined by them are homeomorphic, can we, by certain particularly simple operations, pass from P_1 to a polyhedron P_3 determining the same closed surface as P_1 and having the same planar polygonal schema as P_2? (We know, from what we have already seen, that the closed surfaces determined by P_3 and P_2 are homeomorphic.) In this connection, we are led to the notion of *elementary operations*, which we are going to define.

To an operation performed on a polyhedron corresponds an operation performed on its planar polygonal schema, and conversely. We shall consider four kinds of elementary operations that can be performed on a planar polygonal schema; they are: a one-dimensional subdivision, a one-dimensional union, a two-dimensional subdivision, and a two-dimensional union. A *one-dimensional subdivision* consists in dividing two equivalent sides by means of two equivalent points taken on these two sides respectively. (One thus introduces two new equivalent vertices, and an old pair of equivalent sides becomes two pairs of

equivalent sides. Conveniently oriented arrows are marked on these
new sides.) The reverse process is called a *one-dimensional union*. A *two-dimensional union* consists in joining two polygons into a single polygon
by making two equivalent sides coincide, of which one is on one of the
two polygons and the other is on the other.[1] (One consequently sup-

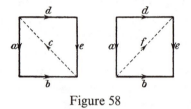

Figure 58

presses a pair of equivalent sides.) The reverse process, which consists
in dividing a polygon in two by means of a diagonal, is called a *two-dimensional subdivision*. (The diagonal then gives a new pair of equivalent sides, on which are marked two conveniently oriented arrows.)

For example, in applying a two-dimensional union to the first two
triangles as well as to the last two of Fig. 57, we obtain the planar polygonal schema of Fig. 58. Fig. 59 shows successively: (1) a one-dimensional union performed on the schema of Fig. 58; (2) a two-dimensional
union; (3) a one-dimensional union. We thus obtain a single polygon of

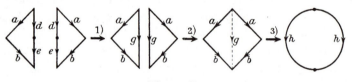

Figure 59

two sides which constitutes, as we have already seen on page 37, the
planar polygonal schema of a polyhedron also obtained from the sphere
(by a polygonal division into a single polygon).

Let P_1 and P_2 be two polyhedra such that the planar polygonal
schema of P_2 is obtained from that of P_1 by an elementary operation,
say, for example, a two-dimensional subdivision. In dividing a face of

---

[1] Of course, it is understood that the placing of two equivalent sides in
coincidence must be performed so that the heads of the arrows on the two sides
come into coincidence.

P_1 into two faces by means of a diagonal, we obtain a polyhedron P_3 having the same schema as P_2. The polyhedra P_3 and P_2 then determine two closed homeomorphic surfaces. Now the polyhedra P_1 and P_3 determine the same closed surface. Therefore the closed surfaces determined by the polyhedra P_1 and P_2 are homeomorphic. It is easily verified that this will be so if the planar polygonal schema of P_2 is obtained from that of P_1 by any finite number of elementary operations. In this case, we shall say that the two polyhedra P_1 and P_2 are *elementarily associated*. Thus *the closed surfaces determined by two elementarily associated polyhedra are homeomorphic*. We have here a sufficient condition for two polyhedra to determine two closed homeomorphic surfaces; we shall later see (page 68) that this condition is also necessary.

23. USE OF NORMAL FORMS OF POLYHEDRA

By the preceding definition, polyhedra can be placed in pairwise disjoint classes in such a way that two polyhedra belong to the same class if and only if they are elementarily associated. To a given polyhedron corresponds a well-determined class, which is composed of all the polyhedra elementarily associated with the given polyhedron, so that a class is theoretically known when one knows a single polyhedron in it. Among all the polyhedra of a class, we are going to try to choose the simplest one possible. This polyhedron is elementarily associated with all the polyhedra of the class and is called the *normal form*. It then becomes easier to find if two polyhedra are elementarily associated, since a necessary and sufficient condition for two polyhedra to be elementarily associated is that they have the same normal form. Moreover, the normal form of a class of polyhedra enables us to gain a general idea of the forms of the corresponding closed surfaces, which are, as we have seen above, all homeomorphic to the closed surface determined by the polyhedron having the normal form.

It is clear that the normal form of a class of elementarily associated polyhedra will depend on the character of simplicity adopted. Among the polyhedra of a class, it is natural, from a certain point of view, to consider as the simplest those for which the numbers of faces, edges, and vertices are as small as possible. The normal forms that we are going to obtain possess this character of simplicity. They each have, in addition, a kind of regularity.

To the normal forms of the polyhedra correspond the normal forms of their planar polygonal schemata. We are going to reduce a polyhedron to its normal form by carrying its planar polygonal schema to normal form.[1]

24. REDUCTION TO NORMAL FORM: I

We shall accomplish the reduction of a planar polygonal schema to its normal form in six stages, exclusively using a finite number of elementary operations in each stage.[2] The aim of the first three stages consists in reducing the number of polygons and the number of nonequivalent sides and vertices. From the fourth stage on, these numbers will no longer change. The aim of the last three stages is to convert the schema obtained at the end of the third stage into another which is as regular as possible. The symbolic representation will facilitate the reduction. Since there are several representations for the same schema, we will take, for each schema, that which will be most convenient for us.

First Stage: If the schema contains more than one polygon, we can first of all, by a finite number of two-dimensional unions, pass to a schema containing only one polygon. This new polygon then has an even number of sides, which are pairwise equivalent. The symbolic representation of the new schema contains only one line, in which each letter appears twice (with or without opposite signs). If this polygon has only two sides, its symbolic representation is necessarily

$$(4) \qquad a^+ a^-$$

or

$$(9) \qquad a^+ a^+ .[3]$$

---

[1] The reader may wish to accept the proof developed in Sections 24 and 25, of the possibility of this reduction, for it is rather tiring to follow. In that case, he can skip to the theorem stated at the end of Section 25.

[2] In the reduction to normal form, we follow the method given by SEIFERT and THRELFALL [22], pages 135–139. For another method of reduction, see J. W. ALEXANDER [1].

[3] There are two other possible symbolic representations: $a^- a^+$ and $a^- a^-$. But $a^+ a^-$ and $a^- a^+$ (or $a^+ a^+$ and $a^- a^-$) are symbolic representations of the same schema.

These are two very simple forms which we shall consider as normal. We shall then suppose, in the following, that the schema is composed of a single polygon and that this polygon has at least four sides.

Second Stage: If, in this polygon, there are two adjacent sides which are equivalent but oppositely oriented, the symbolic representation can be supposed to be of the form

(11) $\sim\!\sim\!\sim a^+ a^- \sim\!\sim\!\sim,$

where the wavy lines $\sim\!\sim\!\sim$ denote the other sides, which we do not need to state explicitly (Fig. 60). Since the polygon has at least four sides, we can suppose, by taking a cyclic permutation of the sides if necessary, that in the symbolic representation (11) there is at least one side before $a^+ a^-$ and at least one side after $a^+ a^-$. Applying a two-dimensional

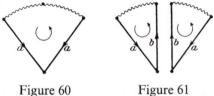

Figure 60   Figure 61

subdivision, we will obtain (if, for example, we mark the new straight and curved arrows as in Fig. 61) two polygons

$$\sim\!\sim\!\sim a^+ b^+ \quad \text{and} \quad b^- a^- \sim\!\sim\!\sim,$$

where b is a new side (Fig. 61). Next, by a one-dimensional union, we form one side c from the two sides a and b, and the polygons become

$$\sim\!\sim\!\sim c^+ \quad \text{and} \quad c^- \sim\!\sim\!\sim$$

(Fig. 62). Finally, we again obtain a single polygon by a two-dimensional union along the side c (Fig. 63).

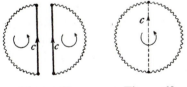

Figure 62   Figure 63

Therefore, if the polygon has at least four sides and if its symbolic representation is of the form (11), we can reduce it to another polygon whose symbolic representation is obtained from (11) by simply suppressing $a^+ a^-$.

By repeating this procedure a finite number of times, we will finally arrive at one of the following two possible cases: either where we will have a polygon of two sides, or where we will have a polygon having at least four sides and not containing two adjacent sides of the form $a^+ a^-$. In the first case, we have already arrived at the normal form (4) or (9). We have to pursue the reduction only for the second case. We will suppose, then, in the following, that the considered schema is composed of a single polygon having at least four sides and that this polygon does not have two adjacent sides of the form $a^+ a^-$.

Third Stage: Let $n \geqq 4$ be the number of sides of such a polygon. It then has n vertices. Denote equivalent vertices by the same letter. If the n vertices of the polygon are all equivalent, the aim of the third stage in our reduction will already have been attained and we will pass on to the following stages. If they are not all equivalent, we shall make a reduction as follows.

Consider a vertex P, the vertices equivalent to P also being denoted by P. By our hypothesis, there exists at least one vertex not equivalent to P. Consequently, there is at least one vertex Q on the perimeter of the polygon which is not equivalent to P and which is the other extremity of a side issuing from a vertex P (Fig. 64). Let a be the side limited by these two vertices and b the adjacent side issuing from P. The other extremity of this side b may be a vertex P, or a vertex Q, or a vertex not equivalent to either P or Q. Join this vertex to Q by a diagonal d. The given polygon is then divided into a triangle $\varDelta$ and a polygon $\varPi$ of $n - 1$ sides (Fig. 64). We now claim that the two adjacent sides a and b of triangle $\varDelta$ are not equivalent. In fact, in the contrary case, we would have $b^+ = a^+$ or $b^+ = a^-$. If $b^+ = a^+$, the vertices P and Q would be equivalent, which is a contradiction; if $b^+ = a^-$, the given polygon would have two adjacent sides of the form $a^+ a^-$, contrary to our hypothesis. Therefore the side equivalent to side b is certainly on the perimeter of polygon $\varPi$ (Fig. 64). Now apply a two-dimensional subdivision along the diagonal d and then a two-dimensional union by identifying the two sides b. We then obtain a polygon which again has n sides (Fig. 65). But, in the new polygon, the number of vertices P is one less than in the given polygon, while the number of vertices Q is increased by one.

There are two possibilities for this new polygon of n sides: either there are two equivalent adjacent sides oriented in opposite senses, or there are not. In the latter case, one can repeat the same procedure to diminish again the number of vertices P. But one will certainly arrive,

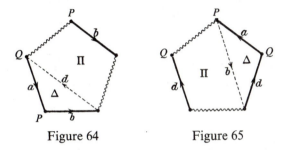

Figure 64   Figure 65

after having repeated this procedure a finite number of times, to a polygon of n sides having adjacent sides of the form a^+a^-. This will happen, at the latest, when the reduced polygon has only one of the vertices P, for, in this case, the two sides having P as a common vertex are necessarily equivalent and oppositely oriented.

When equivalent adjacent sides oriented in opposite senses present themselves, we resort to the process of the second stage. By this process, we will obtain a new polygon: either having only two sides and then already in normal form, or having all its vertices equivalent and the aim of the third stage is attained, or having the number of its sides ≥ 4 and its vertices not all equivalent—then one recommences the procedure.

25. REDUCTION TO NORMAL FORM: II

By the first three stages, which we have already described, we can reduce, with a finite number of elementary operations, each planar polygonal schema to a new schema composed of a single polygon which is of one of the following two types: (1) either it has only two sides; its symbolic representation is then a^+a^- or a^+a^+ and it already has the sought normal form; (2) or it has $n \geq 4$ sides and its n vertices are all equivalent.

Therefore, in the following, we shall concern ourselves exclusively with polygons of the second type.

Fourth Stage: Consider such a polygon. Its sides being pairwise equivalent, each letter representing a side appears twice in its symbolic representation, with or without opposite signs. If, for each pair of equivalent sides, the two signs are opposite, we have nothing to do in the fourth stage and we will pass directly to the fifth stage. If there are two equivalent sides with like signs and if they are not adjacent, we proceed so as to replace them by two equivalent adjacent sides having like signs.

Let b represent two equivalent sides which are not adjacent and which have like signs. The symbolic representation of the polygon is then of the form

$$\sim\!\sim\!\sim b^+ \sim\!\sim\!\sim b^+.$$

Join, by a diagonal a_1, the two vertices toward which the arrows on the two sides are directed (Fig. 66). Next successively apply a two-dimen-

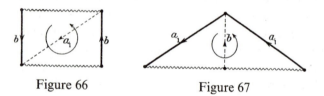

Figure 66   Figure 67

sional subdivision along a_1 and a two-dimensional union along b (Fig. 67). The symbolic representation of the new polygon so obtained is

$$\sim\!\sim\!\sim a_1^+ a_1^+ \sim\!\sim\!\sim.$$

We have thus replaced the two equivalent sides b, which were non-adjacent and with like signs, by two equivalent sides a_1 which are adjacent and also with like signs. And the sides represented by each wavy line $\sim\!\sim\!\sim$ in the original polygon have been carried intact, without breaking, into the new polygon.

If, in this new polygon, there is still a pair of equivalent sides with like signs and which are not adjacent, we can in the same way convert them into adjacent sides $a_2^+ a_2^+$, without disturbing the pair $a_1^+ a_1^+$ already obtained. Continuing in this way, we will stop when each pair of non-adjacent equivalent sides with like signs is replaced by a pair of adjacent equivalent sides with like signs.

If the above exhausts all the sides of the polygon, we will have before us a polygon whose symbolic representation will be of the form

$$\text{(12)} \qquad a_1^+ a_1^+ a_2^+ a_2^+ \cdots a_q^+ a_q^+,$$

a representation sufficiently simple for us to adopt as a normal form.

Fifth Stage: After the fourth stage, we may assume that the considered planar polygonal schema satisfies, in addition to the hypotheses made before the fourth stage,[1] the following conditions: there is at least one pair of equivalent sides having opposite signs; if there are two equivalent sides with like signs, they are adjacent.

Now let c^+ and c^- be a pair of equivalent sides with opposite signs. We claim that there exists at least one other pair of equivalent sides oppositely oriented, say d^+ and d^-, such that the two pairs c^+, c^- and d^+, d^- overlap each other on the perimeter of the polygon (Fig. 68), that is, the symbolic representation of the polygon is of the form[2]

$$\sim\!\!\sim c^+ \sim\!\!\sim d^+ \sim\!\!\sim c^- \sim\!\!\sim d^- \sim\!\!\sim.$$

In fact, if the pair c^+, c^- should not be overlapped by another pair of equivalent sides with opposite signs, the sides between them in the sequence $c^+ \sim\!\!\sim c^-$ would be pairwise equivalent (since we assume that whenever there are two equivalent sides with like signs they are adjacent), and, consequently, the two vertices on the side c are unable to be equivalent, which contradicts our hypothesis (that all the vertices of the polygon are equivalent).

---

[1] Recall that these are the following hypotheses: the schema is composed of a single polygon; this polygon has $n \geq 4$ sides; and its n vertices are all equivalent.

[2] There are other possible forms for the symbolic representation, such as

$$\sim\!\!\sim c^- \sim\!\!\sim d^+ \sim\!\!\sim c^+ \sim\!\!\sim d^- \sim\!\!\sim$$

or

$$\sim\!\!\sim d^- \sim\!\!\sim c^+ \sim\!\!\sim d^+ \sim\!\!\sim c^- \sim\!\!\sim,$$

etc., but each of these can be put in the form

$$\sim\!\!\sim c^+ \sim\!\!\sim d^+ \sim\!\!\sim c^- \sim\!\!\sim d^- \sim\!\!\sim$$

by conveniently changing the sense of arrows and by a cyclic permutation of the sides.

The two overlapping pairs c^+, c^- and d^+, d^- can, by a finite number of elementary operations, be converted into new pairs forming a sequence $a^+b^+a^-b^-$. Figs. 68 through 70, which are self-explanatory, show these operations. It is to be noticed that, during these operations, the sides represented by each wavy line $\sim\!\sim$ in the original polygon have been carried intact, whence pairs of adjacent sides of the form $a_1^+a_1^+$ obtained in the fourth stage are preserved during these operations.

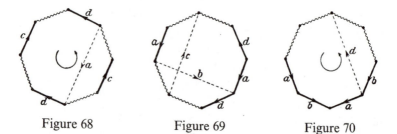

Figure 68   Figure 69   Figure 70

If, in the polygon thus reduced, there are, apart from the pairs a^+, a^- and b^+, b^-, another pair of equivalent sides with opposite signs, say e^+, e^-, this pair is not overlapped by either a^+, a^- or b^+, b^-. There then necessarily exists still another pair f^+, f^- such that the two pairs e^+, e^- and f^+, f^- are overlapping. These two overlapping pairs can similarly be replaced by a sequence of adjacent sides of the same form as $a^+b^+a^-b^-$. In the course of this new reduction, the sequence of adjacent sides already obtained of the form $a_1^+a_1^+$ or the form $a^+b^+a^-b^-$ always remain intact.

Continuing in this way, we will finally arrive at a polygon whose symbolic representation will be uniquely formed of sequences of adjacent sides of the form $a_1^+a_1^+$ or of the form $a^+b^+a^-b^-$.

If, in particular, the symbolic representation contains only sequences of adjacent sides of the form $a^+b^+a^-b^-$, that is, if it is of the form

$$(8) \qquad a_1^+b_1^+a_1^-b_1^-a_2^+b_2^+a_2^-b_2^-\cdots a_p^+b_p^+a_p^-b_p^-,$$

we then already have a simple form which we will consider as normal.

Sixth Stage: It finally remains for us to treat the following case: the symbolic representation of the polygon is formed exclusively of sequences of adjacent sides of the form $a_1^+a_1^+$ or of the form $a^+b^+a^-b^-$, and each of these two forms is actually present. In this case, we can replace each quadruple sequence $a^+ b^+ a^- b^-$ by two sequences of the form $a_1^+ a_1^+$. In fact, let us consider the quadruple sequence $a^+ b^+ a^- b^-$ and a sequence $c^+ c^+$ of two terms. We can first of all convert $a^+ b^+ a^- b^-$ and $c^+ c^+$ into three pairs of equivalent sides such that the signs of both sides of each pair are the same (Figs. 71 and 72). Next, by resorting three

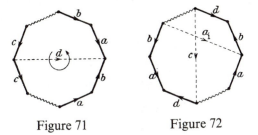

Figure 71   Figure 72

times to the process of the fourth stage, these three pairs can, in turn, be replaced by three new adjacent pairs, each of which is composed of two equivalent sides with like signs (Figs. 72 through 75). We thus see that the case actually leads us finally to a polygon whose symbolic representation is of the normal form (12).

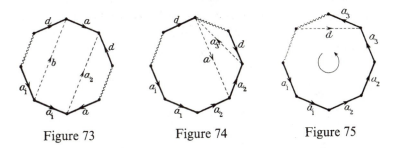

Figure 73   Figure 74   Figure 75

In summary, we have established the purely combinatorial theorem [cc]: *Every polyhedron is elementarily associated with a polyhedron (having only one face) whose symbolic representation is of one of the following types:*

(4) $a^+ a^-$,

(8) $a_1^+ b_1^+ a_1^- b_1^- a_2^+ b_2^+ a_2^- b_2^- \cdots a_p^+ b_p^+ a_p^- b_p^-$, $(p = 1, 2, 3, \cdots)$,

(12) $a_1^+ a_1^+ a_2^+ a_2^+ \cdots a_q^+ a_q^+$, $(q = 1, 2, 3 \cdots)$.

As an illustrative exercise, consider a sphere with three cross caps. We have seen (page 44) that this can be obtained from a region bounded by three circumferences by identifying the diametrically opposite points on each of the three circumferences. This region bounded by three circumferences can be regarded as the union of two polygons, each of six sides (Fig. 76). We thus have a planar polygonal schema composed

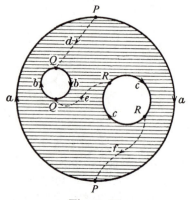

Figure 76

of two polygons each of six sides (Fig. 77). The reader can himself reduce this schema to its normal form, which is of the type (12) with $q = 3$. More generally, one can show that (12) is the normal form of all polyhedra arising from a sphere with q cross caps.

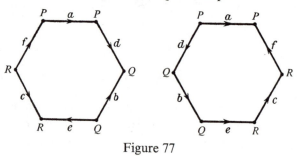

Figure 77

26. CHARACTERISTIC AND ORIENTABILITY

In Chapter Two (pages 27 and 31) we have already spoken of the notions of characteristic and orientability. But, like prior to giving precise definitions of closed surfaces and polyhedra, we there expressed ourselves very vaguely. We now define these two notions more precisely. In the same way that we defined closed surfaces by starting with polyhedra, similarly, to define the characteristic and orientability of closed surfaces, we first define these notions for polyhedra.

Given a polyhedron, let n_v be the number of its vertices, n_e that of its edges, and n_f that of its faces. The number

(13) $$\chi = n_v - n_e + n_f$$

is called the *characteristic* of the polyhedron. If we consider, instead of this polyhedron, its planar polygonal schema, n_v is the number of nonequivalent vertices, n_e is the number of nonequivalent edges, and n_f is the number of polygons. When applying an elementary operation to the planar polygonal schema, the characteristic remains invariant. In fact, a one-dimensional subdivision leaves n_f unaltered, while each of the numbers n_v and n_e is increased by one. A two-dimensional subdivision increases each of the numbers n_e and n_f by one, while n_v remains the same. It follows, then, that if two planar polygonal schemata are elementarily associated, they have the same characteristic. Thus *two elementarily associated polyhedra have the same characteristic.*

A polyhedron is said to be *orientable* if we can determine a cyclic sense of direction for the perimeter of each polygon of its planar polygonal schema so that, in the corresponding symbolic representation, each letter representing two equivalent sides appears twice with opposite signs.[1] When this is impossible, the polyhedron is said to be *nonorientable*. One easily proves that the property of a planar polygonal schema being orientable or being nonorientable is preserved when one performs an elementary operation on the schema, whence *two elementarily associated polyhedra are both orientable or both nonorientable.*

For polyhedra with normal forms (4), (8), and (12), we have, respectively,

(14) $$\chi = 2 - 1 + 1 = 2,$$

(15) $$\chi = 1 - 2p + 1 = 2(1 - p),$$

(16) $$\chi = 1 - q + 1 = 2 - q.$$

---

[1] In particular, for a polyhedron having only one face to be orientable, it is necessary and sufficient that each letter representing two equivalent sides appear twice with opposite signs in every possible symbolic representation of this polyhedron.

Polyhedra represented by (4) and (8) are orientable, whereas those of type (12) are not.[1] From these considerations, bearing in mind that two elementarily associated polyhedra have the same characteristic and are or are not simultaneously orientable, we conclude that, no matter what be the numbers p and q, the formulas (4), (8), and (12) represent polyhedra that are not elementarily associated in pairs, and that to any two distinct values of p (or of q) in (8) [or in (12)] correspond two polyhedra that are not elementarily associated. It follows that the polyhedra represented by (4), (8), and (12) actually constitute distinct normal forms. *Two polyhedra are elementarily associated if and only if they have the same normal form.*

We are now going to verify the following fact (already pointed out on page 56): for any polyhedron P, the numbers n_v, n_e, and n_f are, respectively, at least equal to the corresponding numbers for the normal form of P. In fact, if the normal form of P is (4), we have for P,

$$n_v = n_e - n_f + \chi \geq \chi = 2.^2$$

If the normal form of P is (8), we have

$$n_e = n_v + n_f - \chi \geq 2 - \chi = 2p.$$

If the normal form of P is (12), we have

$$n_e \geq 2 - \chi = q.$$

From (14) and (15) we see that the characteristic of an orientable polyhedron can assume only the even integral values ≤ 2. Similarly, we see from (16) that the characteristic of a nonorientable polyhedron can assume any integral value ≤ 1.

Given a polyhedron, if it is orientable, its normal form is necessarily (4) or (8)—it is (4) if the characteristic is 2 and it is of type (8) if the characteristic < 2, and in the latter case, the characteristic determines the value of p in (8) by the equality (15). For a nonorientable poly-

---

[1] This follows from the preceding footnote.

[2] For any polyhedron, $n_e - n_f \geq 0$. In fact, each of the n_f faces is a polygon of at least two sides; the total number $2n_e$ of sides of these n_f polygons is then $\geq 2n_f$.

hedron, the normal form is necessarily of type (12) and its characteristic determines the value of q in (12) by the equality (16). Thus, if two polyhedra have the same characteristic and are both orientable or both nonorientable, they have the same normal form and consequently are elementarily associated.

We have thus established the following theorem, due to M. Dehn and P. Heegaard: *In order that two polyhedra be elementarily associated, it is necessary and sufficient that they have the same characteristic and be simultaneously orientable or simultaneously nonorientable.*

Since any closed surface can be regarded as a polyhedron, the characteristic of this polyhedron will also be called the characteristic of the closed surface; and the closed surface will be said to be orientable or not according as the polyhedron is orientable or not. But the same closed surface can give rise to different polyhedra by means of different polygonal divisions. Can it be, then, that two polyhedra obtained from the same closed surface have distinct characteristics or be such that one is orientable and the other not? If such situations could exist, the preceding definitions of the characteristic and of the orientability of a closed surface would not be legitimate. But such cannot be: *two polyhedra obtained from the same closed surface by different polygonal divisions always have the same characteristic and are both orientable or both nonorientable.* In other words, the characteristic and the property of being orientable or not are two intrinsic properties of the closed surface; they are defined by means of a polygonal division of the closed surface, but in reality they are independent of it. We accept this fact without proving it (see DE KERÉKJÁRTÓ [14], pages 136, 145).

27. THE PRINCIPAL THEOREM OF THE TOPOLOGY OF CLOSED SURFACES

Consider, now, two closed homeomorphic surfaces. To a polygonal division of one of them corresponds a polygonal division of the other. Thus two closed homeomorphic surfaces can be regarded as two polyhedra having the same planar polygonal schema. These two polyhedra then have the same characteristic and are both orientable or both nonorientable. It follows that two closed homeomorphic surfaces have the same characteristic and are simultaneously orientable or simultaneously nonorientable. Thus, *the characteristic and the orientability of a closed surface are not only independent of the polygonal division: they are, in addition, two topological invariants.*[1]

We are now in a position to prove the following proposition. *In order that two closed surfaces determined by two polyhedra be homeomorphic, it is necessary and sufficient that these polyhedra be elementarily associated.* The sufficiency of this condition has already been established on page 55. Consider, then, two polyhedra which determine two closed homeomorphic surfaces. By the topological invariance of the characteristic and of orientability, these two surfaces, and consequently also the two polyhedra, have the same characteristic and are simultaneously orientable or simultaneously nonorientable. Therefore, by the theorem of Dehn and Heegaard, the two polyhedra are elementarily associated.

Thus, to different types (4), (8), and (12) of nonelementarily associated polyhedra correspond nonhomeomorphic closed surfaces. The polyhedron (4) comes from the *sphere* (see page 57). The polyhedron (8) is obtained from a *generalized torus with p holes* (for the cases $p = 1, 2,$ and 3, see Figs. 4, 36, and 40). The polyhedron (12) is obtained from the *sphere with q cross caps* (for the cases $q = 1$ and 2, see Figs. 25 and 54).

The only topologically distinct (that is, nonhomeomorphic) *types of closed orientable surfaces are the sphere, the torus, and, in general, the generalized torus with p holes* ($p = 1, 2, 3, \cdots$). *For closed nonorientable surfaces, the only topologically distinct types are given by the sphere with q cross caps* ($q = 1, 2, 3, \cdots$).

Since two closed surfaces determined by two polyhedra are homeomorphic if and only if these polyhedra are elementarily associated, we immediately deduce, by applying the theorem of Dehn and Heegaard, the following important proposition, known as *the principal theorem of the topology of closed surfaces: two closed surfaces are homeomorphic if and only if they have the same characteristic and are both orientable or both nonorientable.* The principal problem of the topology of closed surfaces is thus completely solved. And it is seen, in the last theorem, that one no longer interposes the secondary notion of polyhedron.

---

[1] Cf. H. SEIFERT and W. THRELFALL [22], page 140.

28. APPLICATION TO THE GEOMETRIC THEORY OF FUNCTIONS

For those acquainted with Riemann surfaces and algebraic functions, we now mention an important application of the above topological classification of closed surfaces. The Riemann surface of an analytic function plays a fundamental role in the geometric theory of functions. The study of the topological properties of a Riemann surface brings to light the deepest properties of the corresponding function. It is known that the closed Riemann surfaces correspond to algebraic functions. Furthermore, it is proved that every Riemann surface is orientable, and, conversely, every closed orientable surface is homeomorphic to a Riemann surface, so that the topological theory of Riemann surfaces of algebraic functions identifies itself with that of closed orientable surfaces. It follows, then, that the only topological types of Riemann surfaces of algebraic functions are the sphere, the torus, and, in general, the generalized torus with p holes ($p = 1, 2, 3, \cdots$) [dd].

We content ourselves with only these brief indications to show the role played by combinatorial topology in the theory of functions, a theory that at first view appears far removed from topology. The topological study of analytic functions today constitutes a whole new and deep mathematical discipline,[1] which is beyond the scope of this book.

29. GENUS AND CONNECTION NUMBER OF CLOSED ORIENTABLE SURFACES

In this final section we restrict ourselves to closed orientable surfaces. We have seen, by the principal theorem (page 68), that two closed orientable surfaces are homeomorphic if and only if they have the same characteristic, so that the characteristic is a topological invariant that suffices to characterize the different topological types of closed orientable surfaces. We are now going to show that the role played here by the characteristic can also be played by other topological invariants which have greater intuitive meaning than the characteristic.

First of all, let us make these intuitive observations. Every closed Jordan curve on the sphere divides the sphere into two pieces. We see in Fig. 4 that, for the torus, neither the closed Jordan curve a nor the closed Jordan curve b divides the surface. But these two curves a and b have a common point. In fact, it is impossible to draw two disjoint[2] closed

---

[1] The excellent work of S. Stoïlow [25] is devoted to this theory.

[2] That is, having no point in common.

Jordan curves on the torus without dividing the surface. In Fig. 78 we see that one can draw two disjoint closed Jordan curves on the generalized torus of two holes without dividing the surface; but there do not exist three such curves.

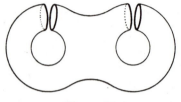

Figure 78

We shall call the maximum number of pairwise disjoint closed Jordan curves that can be drawn on a closed orientable surface without dividing the surface, the *genus* of the surface.[1] By this definition, the genus of the sphere is 0, that of the torus is 1, and, more generally, the generalized torus with p holes has genus p.

For two homeomorphic closed orientable surfaces, it is clear that to a closed Jordan curve on one of them corresponds a closed Jordan curve on the other. If the first curve divides the first surface, the second curve will likewise divide the second surface. Thus two homeomorphic closed orientable surfaces have the same genus. In other words, *the genus is a topological invariant.*

Figure 79

If we do not insist that the closed Jordan curves be disjoint, we can draw two closed Jordan curves on the torus without dividing the surface (Fig. 4), but no more than two. Similarly, there exist four and no more than four closed Jordan curves on the generalized torus with two holes which do not divide the surface (Fig. 79).

---

[1] De Kerékjártó [14], page 146.

We call the maximum number of closed Jordan curves (*disjoint or not*) that can be drawn on a closed orientable surface without dividing the surface, the *connection number* of the surface.[1] It is clear that *the connection number also is a topological invariant*. We have seen that the connection number of the sphere is 0, that of the torus is 2, and that of the generalized torus with two holes is 4. More generally, the connection number of the generalized torus with p holes is $2p$. Thus, for closed orientable surfaces, the genus is equal to exactly half the connection number.

Denoting the genus and the connection number of a closed orientable surface by γ and κ, respectively, we have, by (14) and (15),

$$(17) \qquad\qquad \chi = 2(1 - \gamma),$$

$$(18) \qquad\qquad \kappa = 2 - \chi,$$

so that any two of the three numbers χ, γ, and κ are determined by the third. This leads to the following two propositions, the first of which was found in 1866 by JORDAN [13], and whose geometric content is very strikingly independent of the notion of polygonal division:

A necessary and sufficient condition for two closed orientable surfaces to be homeomorphic is that they have the same genus.

A necessary and sufficient condition for two closed orientable surfaces to be homeomorphic is that they have the same connection number.

---

[1] SIEFERT and THRELFALL [22], page 147.

Bibliography [ee]

1. J. W. ALEXANDER, "Normal forms for one- and two-sided surfaces," *Annals of Math.*, *16* (1915), 158-161.
2. P. ALEXANDROFF and H. HOPF, *Topologie, I.* Berlin: Julius Springer, 1935.
3. L. ANTOINE, "Sur l'homéomorphie de deux figures et de leurs voisinages," *Journ. Math. pures et appl.*, 8ᵉ série, *4* (1921), 221–325.
4. N. BOURBAKI, *Eléments de mathématiques: Livre III, Topologie général* (Actual. scient. et industr., fasc. nº 858). Paris: Hermann, 1940.
5. A. DENJOY, "Démonstration de la propriété fondamentale des courbes de M. Jordan," *C. R. Acad. Sci.*, *167* (1918), 389–391.
6. A. ERRERA, *Du coloriage des cartes et de quelques questions d'Analysis situs.* Paris: Gauthier-Villars, 1921.
7. M. FRÉCHET, *Les espaces abstraits et leur théorie considérée comme introduction à l'analyse générale.* Paris: Gauthier-Villars, 1928.
8. L. GODEAUX, *Les géométries* (Collection Armand Colin, nº 206). Paris: Armand Colin, 1937.
9. P. J. HEAWOOD, "Map colour theorem," *Quarterly Journ. Math.*, *24* (1890), 332–338.
10. L. HEFFTER, "Ueber das Problem der Nachbargebiete," *Math. Annalen*, *38* (1891), 477–508.
11. D. HILBERT and S. COHN-VOSSEN, *Anschauliche Geometrie.* Berlin: Julius Springer, 1932.
12. C. JORDAN, *Cours d'analyse de l'Ecole Polytechnique, I.* Paris: Gauthier-Villars, 1893.
13. C. JORDAN, "Sur la déformation des surfaces," *Journ. Math. pures et appl.*, 2ᵉ série, *11* (1866), 105–109.
14. B. M. DE KERÉKJÁRTÓ, *Vorlesungen über Topologie, I.* Berlin: Julius Springer, 1923.

73

15. B. M. DE KERÉKJÁRTÓ, "Démonstration élémentaire du théorème de Jordan sur les courbes planes," *Acta Sci. Math. Szeged, 5* (1930), 56–59.

16. W. KILLING and H. HOVESTADT, *Handbuch des mathematischen Unterrichts, II.* Leipzig-Berlin: Teubner, 1913.

17. H. LEBESGUE, "Quelques conséquences simples de la formule d'Euler," *Journ. Math. pures et appl.*, 9ᵉ série, *19* (1940), 27–43.

18. F. LEVI, *Geometrische Konfigurationen.* Leipzig: Hirzel, 1929.

19. H. POINCARÉ, *Dernières pensées* (Bibliothèque Philos. scient.). Paris: Ernest Flammarion, 1913.

20. A. SAINTE-LAGUË, *Géométrie de situation et jeux* (Mémorial Sci. Math., fasc. nº 41). Paris: Gauthier-Villars, 1929.

21. E. SCHMIDT, "Ueber den Jordanschen Kurvensatz," *Sitzsber. Preuss. Akad. Wiss.*, *28* (1923), 318–329.

22. H. SEIFERT and W. THRELFALL, *Lehrbuch der Topologie.* Leipzig: Teubner, 1934.

23. F. SEVERI, *Topologia* (Faculted de Ciencias exactas, físicas y naturales, Serie B, Publicación nº 9). Buenos-Aires, 1931.

24. W. SIERPIŃSKI, *Introduction to General Topology.* Toronto: The Univ. of Toronto Press, 1934.

25. S. STOÏLOW, *Leçons sur les principes topologiques de la théorie des fonctions analytiques.* Paris: Gauthier-Villars, 1938.

26. H. TIETZE, "Einige Bemerkungen über das Problem des Kartenfärbens auf einseitigen Flächen," *Jahresber. Deutsch. Math. Vereinigung, 19* (1910), 155–159.

27. G. T. WHYBURN, *Analytic Topology* (Amer. Math. Soc. Colloq. Publications, no. 28). New York: American Mathematical Society, 1942.

28. R. L. WILDER, "Point sets in three and higher dimensions and their investigation by means of a unified analysis situs," *Bull. Amer. Math. Soc.*, *38* (1932), 649–692.

Translator's Notes

[a] *Section 1.* The portion of a plane bounded by two concentric circles in the plane is called an *annulus*.

[b] *Section 1.* For many centuries geometers accepted as intuitively obvious the fact that a simple closed curve Σ in a plane π divides the rest of π into two sets in such a way that any two points in the same set can be joined by a polygonal path in π not intersecting Σ, whereas two points in different sets cannot be so joined. This fact certainly does seem obvious if Σ is a circle or a convex planar polygon, but Figs. 80 and 81 show that the fact is not so obvious for more complicated simple closed planar curves, and, of course, it is possible to draw much more involved simple closed planar curves than those of Figs. 80 and 81.

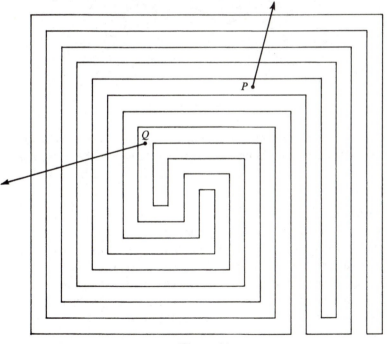

Figure 80

It was not until the second half of the last century that it was thought that a proof of the concerned fact was necessary. Surprisingly, such a proof was not easily found and the problem challenged mathematicians for a number of years. The first successful proof was given by Camille Jordan in his famous *Cours d'analyse* [12], which accounts for the connection of his name with the theorem. Though today many proofs exist, and some recent ones are considerably simpler than Jordan's original

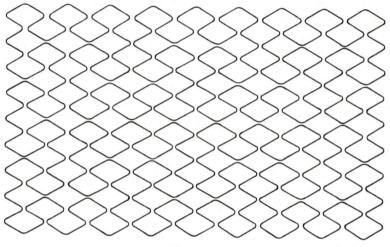

Figure 81

proof, they are all too long or too complex to present here. If, however, we restrict ourselves to simple closed *polygonal* curves in the plane, then a relatively easy solution is possible, and we here indicate such a solution.

THEOREM. *If Σ is any simple closed polygon in a plane π, the points of π which are not on Σ can be divided into two sets in such a way that any two points in the same set can be joined by a polygonal path in π not intersecting Σ, whereas two points in different sets cannot be so joined.*

Choose a direction R in the plane π which is not parallel to any side of the polygon Σ. For each point P of π not on Σ denote by $R(P)$ the ray, or half-line, starting at P and having the direction R, and denote by $\sigma(P)$ the number of points of intersection of $R(P)$ with Σ, where, if $R(P)$ passes through a vertex of Σ, the intersection with Σ is counted if and only if Σ

actually crosses $R(P)$ at the concerned vertex. (For example, in Fig. 82, $\sigma(U) = 2$ and $\sigma(V) = 1$.) Let A be the set of all points P of π, not on Σ, for which $\sigma(P)$ is even, and let B be the set of all points P of π, not on Σ, for which $\sigma(P)$ is odd.

Clearly, if a point P moves along a line segment which does not intersect Σ, the number $\sigma(P)$ can change only when $R(P)$ moves across a

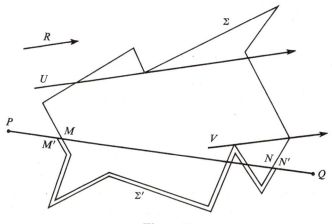

Figure 82

vertex of Σ at which Σ does not cross $R(P)$, and then the change in $\sigma(P)$ is necessarily even. It follows that the points on any line segment, and hence on any polygonal path, in π not intersecting Σ are all in A or all in B. This proves that no point of A can be joined by a polygonal path in π to a point of B without intersecting Σ.

It remains to show that if P and Q are any two points both in set A or both in set B, then P and Q can be joined by a polygonal path in π lying entirely in that set. Consider (see Fig. 82) the line segment PQ. If this line segment does not intersect Σ, then the segment itself is a satisfactory path. If the line segment PQ does intersect Σ, let M be that point of intersection of PQ with Σ nearest P, and N that intersection of PQ with Σ nearest Q. Now construct a polygonal path Σ' from P to Q as follows. Go along the segment PM to a point M' between P and M and very near to M, then along a polygonal path, beginning at M', with sides close to those of Σ but never crossing Σ, until a point N' on PQ very near to N is reached, and finally along the segment $N'Q$ to Q. Now N' must lie between N and Q. For otherwise we would have N between N' and Q and $\sigma(N')$ and $\sigma(Q)$ would differ by 1 and N' and Q would belong to different sets A and B. But N' belongs to the same set as does P (by the construction of the polygonal path Σ') and Q belongs to the same set as does P (by assumption). This contradiction proves that N is not between N' and Q, whence we must have N' between N and Q. We now see that the described polygonal path Σ' joining P and Q lies entirely in set A or entirely in set B, and our theorem is established.

Since a simple closed polygon in π is bounded, there are points in π having σ equal to zero. These belong to set A. Consequently, we call set A the *exterior* of polygon Σ and set B the *interior* of polygon Σ. Thus a point P of π is interior to Σ if and only if P is not on Σ and a ray $R(P)$ not along any side-line of Σ and not passing through any vertex of Σ cuts Σ in an *odd* number of points. For example, point P in Fig. 80 is an interior point and point Q is an exterior point.[1]

[c]   *Section 2.*   Recent research into the history of the four-color conjecture[2] has corrected certain traditional and unfounded accounts. For instance, it has been reported that the conjecture had its origin in a puzzle communicated to A. F. Möbius by his friend Weiske, and that Möbius then mentioned the puzzle to his class in his lectures of 1840. The puzzle amounted to showing that it is impossible to have five regions each having a common boundary with every other region. It seems certain, now, that this puzzle had no historical link with the origin of the four-color conjecture. It has also been reported that cartographers had been aware of the conjecture long before it was made public, but there seems to be no real evidence to support this statement.

Rather than being the culmination of a series of individual efforts, it now appears that the four-color conjecture flashed across the mind of Francis Guthrie sometime around 1850, when he was a mathematics

---

[1] The problem of recognizing the "inside" and the "outside" of a simple closed planar curve recalls the story of the drunk man observed one day walking round and round a great stone pillar of a large building and moaning to himself. When asked by a passerby why he was so sad, he replied, "Look, I'm walled in."

[2] See Kenneth O. May, "The origin of the four-color conjecture," *Isis, 56*, 3, No. 185 (1956), 346–348.

graduate, while coloring a map of England. He attempted a proof of the conjecture but considered his effort as unsatisfactory. Later, in October of 1852, his brother Frederick Guthrie, while attending a class of Professor Augustus De Morgan, communicated the problem, but not the attempted proof, to his teacher. De Morgan gave the problem some attention and tried without success to interest other mathematicians in attempting a solution. Finally the problem reached the attention of Professor Arthur Cayley, who, on June 13, 1878, at a meeting of the London Mathematical Society, announced that he had been unable to obtain a general proof of the conjecture.

In the following year, in the *American Journal of Mathematics*, A. B. Kempe, a British barrister-at-law, published a "proof" of the conjecture. A simplified version of this "proof" was published later in the same year in the *Transactions of the London Mathematical Society*, and again in the following year, 1880, in *Nature*. In 1880, in the *Proceedings of the Royal Society of Edinburgh*, appeared another proof, by P. G. Tait, but this proof was based upon an unproved assumption concerning a closed network of arcs. Then, in 1890, in an article in the London *Quarterly Journal of Mathematics*, P. J. Heawood pointed out a flaw in Kempe's reasoning. This flaw has never been circumvented, and the four-color conjecture remains one of the outstanding unverified conjectures in mathematics today.

Heawood's work was far from being merely destructive in nature. In addition to simplifying some of Kempe's work, Heawood established the seven-color theorem: *to color any map on a torus requires at most seven colors*. Heawood has the honor of being that mathematician who has spent more time and energy than any other individual on the map-coloring problem. A formula yielding a sufficient number of colors was established by Heawood for all closed surfaces of Euler characteristic less than 2. That this number is also the necessary number has been shown for infinitely many orientable and nonorientable closed surfaces. But Philip Franklin, in the *Journal of Mathematics and Physics* of 1934, established the remarkable theorem that for a Klein bottle, for which the Heawood sufficient number is seven, six colors actually suffice. For a sphere, the one case for which Heawood's formula has not been established, five colors have been shown to be sufficient. If Heawood's formula should be assumed to hold for a sphere, this sufficient number would be four. No one has yet produced a spherical map requiring five colors.

Many subsequent mathematicians (and laymen too) have worked on the four-color conjecture and on map-coloring in general, and their results have added greatly to our insight into the problem and have placed increasingly greater restrictions on a spherical map that may need five colors. In 1920, Philip Franklin showed that all maps of 25 or fewer regions can be colored with four colors. In 1926, C. N. Reynolds raised this number to 27; then Franklin, in 1936, raised it to 31; and C. E. Winn, in 1940, raised it to 35.

It is taunting that though the color problem has been solved for many complicated surfaces, it still remains unsolved for the simplest surface of all—the sphere.

[d]   *Section 2.*   The boundaries are assumed to be simple closed curves.

[e]   *Section 2.*   In this note we shall prove that five colors are sufficient to color any map on a sphere. We shall find it convenient to distinguish between a map that covers an entire closed surface and a map that may cover only a part of the surface. We shall call the former a map *of* the surface and the latter a map *on* the surface. Clearly, if n colors suffice to color any map of a given closed surface, then n colors will suffice to color any map on the surface, because a map on the surface is either a map of the surface or part of a map of the surface. We commence by establishing a useful lemma. For the lemma and its proof we let v, e, and f denote the number of vertices, edges, and faces (regions) of a map of a closed surface of Euler characteristic χ, and denote by a the average number of edges per face for the map. Since $af = 2e$, we note that a is not necessarily an integer. We use the fact, established in Section 26, that for any map *of* a closed surface of Euler characteristic χ, $v - e + f = \chi$.

LEMMA.   *For any map of a closed surface of Euler characteristic χ, $a \leqq 6(1 - \chi/f)$.*

There is no loss of generality in assuming that at least three edges radiate from each vertex, for a vertex from which only two edges radiate can be suppressed by amalgamating the two edges into one longer edge. We then have $3v \leqq 2e = af$. Therefore, since $v - e + f = \chi$,

$$e \leqq 3(e - v) = 3(f - \chi),$$

whence

$$a = 2e/f \leqq 6(1 - \chi/f).$$

THEOREM. *Five colors are sufficient to color any map on a sphere.*

We shall employ mathematical induction. Five colors are certainly sufficient if the number of faces does not exceed five. Assume five colors

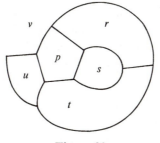

Figure 83

are sufficient for maps of a sphere having $f - 1$ or fewer faces and consider any map M of the sphere having f faces. Since, by the above lemma, $a < 6$, there is at least one face p of M having five or fewer edges. If p has fewer than five edges, then contracting p to a point O yields a map M_1 of the sphere having $f - 1$ faces, which, by our inductive hypothesis, can be colored with five colors. Since there are at most four colors about the point O, when we restore face p, we have a color left to color it, and M is colorable with five colors. If face p has five edges, let r, s, t, u, and v (see Fig. 83) denote the consecutive faces adjacent to p. Now either r and t have no boundary in common, or s and u have no boundary in common. Adjust the notation so that s and u have no boundary in common. By effacing the edges between p and s and between p and u, we obtain a map M_1 of the sphere having fewer faces, which, by our inductive hypothesis, can be colored with five colors. But if M_1 can be colored with five colors, then M can also be colored with five colors, for our construction guarantees that upon restoring the pentagon p, the pair of faces s and u adjacent to p have the same color;

consequently, p is bounded by at most four colors and a fifth color is left to color p.

[f] *Section 2.* Most mathematicians who have worked on the four-color conjecture believe it to be true, but some feel it to be false and say that the difficulty of proving it false lies in the fact that a spherical map requiring five colors probably must contain so many regions that such a map is very hard to find and to test. Many reduction schemes have been devised and many equivalent forms of the conjecture have been made.[*]

[g] *Section 2.* In this note we shall establish Heawood's formula yielding a sufficient number of colors to color any map on a closed surface of Euler characteristic $\chi < 2$. We first prove a preliminary theorem and introduce a convenient notation.

THEOREM 1. *Six colors are sufficient to color any map on a closed surface of Euler characteristic $\chi > 0$.*

We shall employ mathematical induction. Six colors are certainly sufficient if the number of faces (regions) does not exceed six. Assume six colors are sufficient for maps of the surface having $f - 1$ faces and consider any map M of the surface having f faces. Since $\chi > 0$, the lemma of Note [e] guarantees that the average number of edges per face for M is less than six. Thus there is at least one face of M having five or fewer edges. Contracting such a face to a point O yields a map M_1 of the surface having $f - 1$ faces, which, by our inductive hypothesis, can be colored with six colors. Since there are at most five colors about the point O, when we restore the contracted face, we have a sixth color left to color it.

NOTATION. We shall denote by n_χ the real number

$$(7 + \sqrt{49 - 24\chi})/2,$$

and by $[n_\chi]$ the greatest integer not exceeding n_χ.

THEOREM 2. $[n_\chi]$ *colors are sufficient to color any map on a closed surface of Euler characteristic $\chi < 2$.*

The case $\chi = 1$ has already been proved in Theorem 1. Therefore we henceforth assume that $\chi \leq 0$. We shall employ mathematical induc-

---

[*] In the summer of 1976, Kennith Appel and Wolfgang Haken of the University of Illinois established the four-color conjecture by an immensely intricate computer-based analysis.

tion. Certainly $[n_\chi]$ colors are sufficient if the number of faces does not exceed n_χ. Hence let M be a map of the surface having $f > n_\chi$ faces, and assume $[n_\chi]$ colors are sufficient for maps of the surface having $f - 1$ faces. Now, from the definition of n_χ, we have

$$n_\chi^2 - 7n_\chi + 6\chi = 0$$

or

$$6(1 - \chi/n_\chi) = n_\chi - 1.$$

Employing the inequality of the lemma of Note [d], we then have

$$a \leqq 6(1 - \chi/f) \leqq 6(1 - \chi/n_\chi) = n_\chi - 1.$$

It follows that there is at least one face of M having $[n_\chi] - 1$ or fewer edges. Contracting such a face to a point O yields a map M_1 of the surface having $f - 1$ faces, which, by our inductive hypothesis, can be colored with $[n_\chi]$ colors. Since there are at most $[n_\chi] - 1$ colors about the point O, when we restore the contracted face, we have a color left to color it.

Theorems 1 and 2 furnish us a number of colors sufficient for coloring maps on any closed surface. These theorems were established by Heawood in 1890, and the expression for $[n_\chi]$ is known as *Heawood's formula*. It is interesting that for the sphere, where $\chi = 2$ and the only case not covered by Theorem 2, the Heawood formula yields the conjectured value 4.

It is possible that there are closed surfaces of Euler characteristic χ on which every map can be colored with fewer than $[n_\chi]$ colors. For many values of χ, however, it has been shown that on the corresponding closed surfaces there exist maps having exactly $[n_\chi]$ mutually contiguous faces, and the color problem is thus solved for these surfaces. The following table shows the values of $[n_\chi]$ for the first few values of χ:

χ	2	1	0	-1	-2	-3	-4	-5	-6	-7	-8	-9	-10
$[n_\chi]$	4	6	7	7	8	9	9	10	10	10	11	11	12

For the torus, the orientable closed surface with $\chi = 0$, Heawood and others have drawn maps having $[n_\chi]$ mutually adjacent faces. L. Heffter has done the same for many of the other orientable closed surfaces, and I. N. Kagno and H. S. M. Coxeter have done this for a number of the nonorientable closed surfaces. In 1959, G. Ringel proved that the Klein bottle is the only nonorientable closed surface not needing as many as $[n_\chi]$ colors, and he showed that the number of colors needed for orientable closed surfaces can never differ from $[n_\chi]$ by more than 2.

Fig. 5 of Section 2 shows a map of the torus having seven mutually adjacent faces, and therefore requiring seven colors to color it. This map, along with Theorem 2, establishes the following theorem.

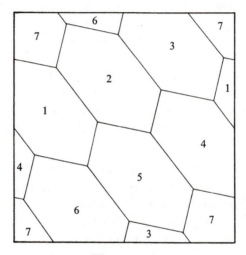

Figure 84

HEAWOOD'S THEOREM.   *Seven is the least number of colors needed to color all maps on a torus.*

Figs. 84, 85, and 86 show other maps of the torus having seven mutually adjacent faces. Fig. 84 is the map constructed by Heawood; Fig. 85 was found by John Leech in 1953; Fig. 86 represents the top and bottom views of a torus.

[h]   *Section 3.*   The regions are assumed to be bounded by simple closed curves.

[i]   *Section 3.*   In this note we show that the greatest possible number of mutually adjacent regions on a closed surface is equal to the

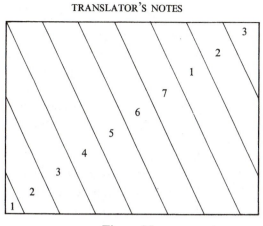

Figure 85

greatest possible number of points on the surface that can be pairwise joined by mutually nonintersecting simple arcs on the surface. We then show that for the plane (or sphere) these two maximum numbers are 4.

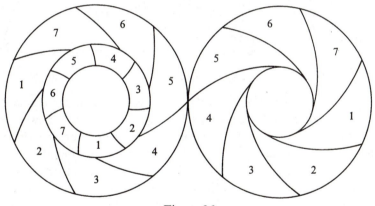

Figure 86

THEOREM 1. *The maximum number of mutually adjacent regions on a closed surface is equal to the maximum number of points on the surface that can be pairwise joined by mutually nonintersecting simple arcs on the surface.*

Let us be given *n* mutually adjacent regions on the surface (see Fig. 87, which shows four mutually adjacent regions). Choose *n* points, one in the interior of each of the *n* adjacent regions. Since two adjacent regions have at least one boundary in common, any two of these points may be connected by a simple arc on the surface and belonging entirely to the two regions containing the two points. Furthermore, the arcs thus drawn can be chosen so that portions lying in any one given region do not intersect, for in this region we need only to connect the chosen interior point with a number of points on its boundary. It follows that the n chosen points are pairwise connected by mutually nonintersecting simple arcs on the surface. Thus the maximum number of points on the surface that can be pairwise joined by mutually nonintersecting simple arcs on the surface is at least equal to the maximum number of mutually adjacent regions on the surface.

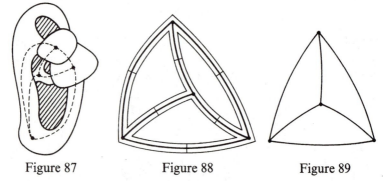

Figure 87   Figure 88   Figure 89

Now let us be given n points on the surface pairwise joined by mutually nonintersecting simple arcs on the surface (see Fig. 88, which shows four such points). Replace each of the connecting arcs by a narrow strip on the surface by adjoining to the points of each arc points on the surface close to and on each side of the arc. Now cut each of these narrow strips in two by a small arc crosswise of the strip. In this way we obtain n star-shaped regions on the surface, each of which borders on every other one. It follows that the maximum number of mutually adjacent regions on the surface is at least equal to the maximum number of points on the surface that can be pairwise joined by mutually nonintersecting simple arcs on the surface.

We conclude that the two concerned maximum numbers are equal.

THEOREM 2. *The maximum number of points in a plane (or on a sphere) that can be pairwise joined by mutually nonintersecting simple arcs in the plane (or on the sphere) is 4.*

Suppose there are five points in a plane (or on a sphere) that are joined in the desired fashion. Then these points and their joins form a map M of the plane (or of the sphere), for which $v = 5$ and $e = 10$. Since, for a map of a plane (or of a sphere), $v - e + f = 2$, we see that M must contain seven faces (regions). Now each face has at least three edges. This means that, for the seven faces, we must have $e \geq 7(3)/2$, which contradicts the fact that $e = 10$. It follows that it is impossible to have five points in a plane (or on a sphere) that are joined in the desired fashion. Since (see Fig. 89) we can have four points in a plane (or on a sphere) joined in the desired fashion, we see that the sought maximum number is 4.

THEOREM 3. *The maximum number of mutually adjacent regions in a plane (or on a sphere) is 4.*

This is now a consequence of Theorems 1 and 2.

In passing, we might mention that the problem of finding the maximum number of points on a closed surface that can be pairwise joined by mutually nonintersecting simple arcs on the surface is sometimes called the *thread problem*.

Allied to the thread problem for the plane is the familiar puzzle of the three houses and the three wells. In this puzzle there are three houses H_1, H_2, H_3 and three wells W_1, W_2, W_3 (see Fig. 90). We are to join each house to each well, but we must not allow any pipe to cross any other pipe. In mathematical language, we are to join each of the three points H_1, H_2, H_3 to each of the points W_1, W_2, W_3 by mutually nonintersecting simple arcs in the plane.

We give an indirect proof of the impossibility of solving the "three houses and three wells" puzzle. Suppose there is a solution to the puzzle. Then the six points and their joins yield a map M of the plane for which $v = 6$ and $e = 9$. Since for any map of a plane, $v - e + f = 2$, we see that M must contain five faces (regions). Now a study of the figure shows that there are no two-sided or three-sided faces, whence each face is bounded by at least four edges. This means that, for the five faces, we must have $e \geq 5(4)/2 = 10$, which contradicts the fact that $e = 9$. It follows that a configuration in the plane of the desired sort cannot exist and the "three houses and three wells" puzzle cannot be solved on a plane.

In 1930, K. Kuratowski proved that any network of mutually non-intersecting simple arcs that cannot be drawn in a plane must contain either the network of Fig. 90 or that of Fig. 91, where only the heavy dots indicate vertices. The reader might like to show that each of these networks can exist on the torus.

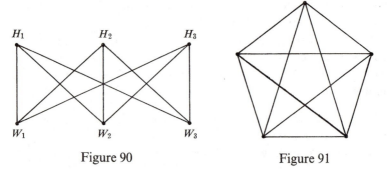

H_1 H_2 H_3

W_1 W_2 W_3

Figure 90   Figure 91

[j]   *Section 5.*   A biunique correspondence is also called a *one-to-one correspondence.*

[k]   *Section 5.*   Thus the frequent description of the intrinsic topology of surfaces as *rubber-sheet geometry* is not accurate. But if we label a material as *pelastic* if it is both deformable and self-penetrable, then we may describe the intrinsic topology of surfaces as *pelastic-sheet geometry.* Fig. 92 shows, for example, how a pelastic torus can be continuously deformed into a knotted torus.

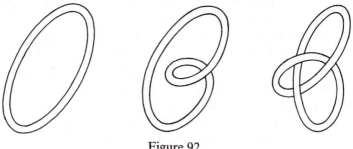

Figure 92

[l]   *Section 5.*   Since the surface of a torus and the surface of a coffee cup are homeomorphic, the hole formed by the handle of the cup corresponding to the hole in the torus, a topologist has been facetiously

described as a person who does not know the difference between a doughnut and a coffee cup.

[m] *Section 5.* Let E_1 denote the set of all points on the unit segment $0 < x \leq 1$ and E_2 the set of all points in the unit square $0 < x,y \leq 1$. A point Z of E_1 may be represented by an unending decimal $z = 0.z_1 z_2 z_3 \cdots$, and a point P of E_2 may be represented by an ordered pair of unending decimals

$$(x = 0.x_1 x_2 x_3 \cdots, y = 0.y_1 y_2 y_3 \cdots).$$

Suppose we let each z_i, x_i, and y_i in these representations denote either a nonzero digit or a nonzero digit preceded by a possible block of zeros. For example, if $z = 0.73028007 \cdots$, then $z_1 = 7$, $z_2 = 3$, $z_3 = 02$, $z_4 = 8$, $z_5 = 007$, $\cdots$. Then it is easy to show that a biunique correspondence is set up between the points of E_1 and those of E_2 by associating with the point $0.z_1 z_2 z_3 \cdots$ of E_1 the point

$$(0.z_1 z_3 z_5 \cdots, 0.z_2 z_4 z_6 \cdots)$$

of E_2, and with the point

$$(0.x_1 x_2 x_3 \cdots, 0.y_1 y_2 y_3 \cdots)$$

of E_2 the point $0.x_1 y_1 x_2 y_2 x_3 y_3 \cdots$ of E_1.

[n] *Section 6.* A relation connecting two elements of a set is called a *dyadic relation*. If R is a dyadic relation connecting elements a and b of a set E, we write $a R b$, and say "a is R-related to b." A dyadic relation R is said to be *reflexive* in a set E if $a R a$ for each element a of E. The dyadic relation R is said to be *symmetric* in E if whenever $a R b$ for a and b in E, we also have $b R a$. The dyadic relation R is said to be *transitive* in E if whenever $a R b$ and $b R c$, for a, b, c in E, we also have $a R c$. A dyadic relation R is called an *equivalence relation* in a set E if R is reflexive, symmetric, and transitive in E. An equivalence relation in a set E partitions E into a collection of mutually exhaustive and nonoverlapping subsets such that all elements of E in a common subset are R-related to each other, and any two elements of E in different subsets are not R-related to each other.

[o] *Section 7.* In 1872, upon appointment to the Philosophical Faculty and the Senate of the University of Erlanger, Felix Klein (1849–1925) delivered, according to custom, an inaugural address in his area of speciality. The address, based upon the work by himself and Sophus Lie (1842–1899) in group theory, set forth a remarkable definition of "a geometry," which served to codify essentially all the existing geometries of the time and pointed the way to new and fruitful avenues of research in geometry. This address, with the program of geometrical study advocated by it, has become known as the *Erlanger Programm.*

In this rather lengthy note, we shall endeavor to elucidate Klein's concept, and in doing so we shall expand upon some of the material of Sections 5 and 7. We commence by stating the definition of a group.

A *group*, which is one of the simplest algebraic structures of consequence, is a nonempty set G of elements in which a binary operation $*$ is defined satisfying the following four postulates:

G1: *For all a, b in G, $a * b$ is in G.*

G2: *For all a, b, c in G, $(a * b) * c = a * (b * c)$.*

G3: *There exists an element i of G such that, for all a in G, $a * i = a$.* (The element i is called an *identity element* of the group. It can be proved that a group possesses only one identity element.)

G4: *For each element a of G there exists an element a^{-1} of G such that $a * a^{-1} = i$.* (The element a^{-1} is called an *inverse element* of a. It can be proved that an element a of a group possesses only one inverse element.)

Klein's application of groups to geometry depends upon the concept of a *biunique transformation* of a set S of elements onto itself, by which (the reader will recall) we simply mean a correspondence under which each element of S corresponds to a unique element of S, and each element of S is the correspondent of a unique element of S. If, in a biunique transformation T of set S onto itself, element a of S corresponds to element b of S, we say that, under the transformation T, a is *carried into* b, or a is *mapped onto* b.

By the *product*, $T_2 T_1$, of two biunique transformations T_1 and T_2 of a set S of elements onto itself, we mean the resultant biunique transformation obtained by first performing transformation T_1 and then transformation T_2. The product of two biunique transformations of a set S onto itself is not necessarily commutative, as is instanced by taking T_1 to be a translation of a distance of one unit in the direction of the positive x-axis applied to the set S of all points in the (x,y)-plane, and T_2 a counterclockwise rotation of the set S through 90° about the origin of coordinates. Under T_2T_1 the point $(1,0)$ is carried into the point $(0,2)$, whereas under T_1T_2 it is carried into the point $(1,1)$. But a product of biunique transformations of a set S onto itself is associative, for if T_1, T_2, and T_3 are any three biunique transformations of a set S onto itself, $T_3(T_2T_1)$ and $(T_3T_2)T_1$ both denote the resultant transformation obtained by first performing T_1, then T_2, then T_3. This can be seen by following the fate, under these transformations, of some arbitrary element a of S. Thus suppose T_1 carries element a into element b, T_2 carries element b into element c, and T_3 carries element c into element d. Then T_2T_1 carries element a directly into element c and T_3 carries element c into element d, whence $T_3(T_2T_1)$ carries element a directly into element d. On the other hand, T_1 carries element a into element b and T_3T_2 carries element b directly into element d, whence $(T_3T_2)T_1$ also carries element a directly into element d.

Let T be a biunique transformation of a set S onto itself which carries each element a of S into its corresponding element b of S. The transformation which undoes transformation T, by carrying each element b of S back into its original element a of S, is clearly biunique; it is called the *inverse* of transformation T and is denoted by T^{-1}. The product of T and T^{-1} is a biunique transformation which clearly leaves all elements of S unchanged; such a transformation is called an *identical transformation*, and will be denoted by I. We note that $TI = T$ for all T.

We may now prove the following important theorem.

THEOREM. *A nonempty set Γ of biunique transformations of a set S onto itself constitutes a group under multiplication of transformations if* (1) *the product of any two transformations of the set Γ is in the set Γ, and* (2) *the inverse of any transformation of the set Γ is in the set Γ.*

To establish this theorem, we first note that, since the product of any two transformations in the set Γ is in the set Γ, Γ is closed under the binary operation of multiplication. This binary operation is associative, as we have pointed out above. Also, if T is a transformation in Γ, then T^{-1} is in Γ. But, as given above, $TT^{-1} = I$ and $TI = T$ for all T of Γ.

All four postulates for a group are thus satisfied, and the theorem is established. Such a group of biunique transformations is briefly referred to as a *transformation group*.

We are now ready to give Klein's famous definition of a geometry: A *geometry* is the study of those properties of a set S which remain invariant when the elements of set S are subjected to the transformations of some transformation group Γ.

To illustrate Klein's definition of a geometry, let S be the set of all points of an ordinary plane, and consider the set Γ of all transformations of S compounded from translations, rotations, and reflections in lines. Since the product of any two such transformations and the inverse of any such transformation are also such transformations, it follows that Γ is a transformation group. The resulting geometry is ordinary *plane Euclidean metric geometry*. Since such properties as length, area, congruence, parallelism, perpendicularity, similarity of figures, collinearity of points, and concurrence of lines are invariant under the group Γ, these properties are studied in plane Euclidean metric geometry.

If, now, the above group Γ is enlarged by including, together with the translations, rotations, and reflections in lines, the homothety transformations (in which each point P is carried into a point P' such that $AP = k(AP')$, where A is some fixed point, k is some fixed positive constant, and A, P, and P' are collinear), we obtain *plane similarity*, or *plane equiform*, *geometry*. Under this enlarged group such properties as length, area, and congruence no longer remain invariant, and hence are no longer subjects of study, but parallelism, perpendicularity, similarity of figures, collinearity of points, and concurrence of lines are still invariant properties, and hence do constitute subject matter for study in this geometry.

Considered from Klein's point of view, plane projective geometry is the study of those properties of the set of points of an extended plane (an ordinary plane augmented by a "line" of ideal points at infinity) which remain invariant when the plane is subjected to the so-called projective transformations, a projective transformation being a product of perspectivities. It is clear that the product of two projective transformations is again a projective transformation, and the inverse of a projective transformation is also a projective transformation. It follows that the set of all projective transformations of an extended plane constitutes a transformation group. Of the previously mentioned properties, only collinearity of points and concurrence of lines still remain invariant.

In all of the above geometries, the fundamental elements upon which the transformations of some transformation group are made to act are points; hence the above geometries are all examples of so-called point geometries. There are, as one might expect, geometries in which entities other than points are chosen for fundamental elements. Thus geometers have studied line geometries, circle geometries, sphere geometries, and various other geometries. In building up a geometry one is at liberty to choose, first of all, the fundamental element of the geometry (point, line, circle, etc.); next, the manifold or space of these elements (plane of points, ordinary space of points, spherical surface of points, plane of lines, pencil of circles, etc.); and finally, the group of transformations to which the manifold of fundamental elements is to be subjected. The construction of a new geometry becomes, in this way, a rather simple matter.

Another interesting feature is the way in which some geometries embrace others. Thus, since the transformation group of plane Euclidean metric geometry is a subgroup of the transformation group of plane equiform geometry, it follows that any theorem holding in the latter geometry must hold in the former. As a further illustration, our theorem above assures us that the set of all projective transformations of an extended plane that carry some fixed line of the plane onto itself constitutes a transformation group; the study of properties of figures of an extended plane that are invariant under the transformations of this group is known as *plane affine geometry*. Again, our theorem assures us that the set of all projective transformations of an extended plane that carry a fixed line onto itself and a fixed point not on the line onto itself also constitutes a transformation group; the study of properties of figures of an extended plane that are invariant under the transformations of this group is known as *plane centro-affine geometry*. The transformation group of plane centro-affine geometry is a subgroup of the transformation group of plane affine geometry, which, in turn, is a subgroup of the transformation group of plane projective geometry. Until recent times, the transformation group of projective geometry contained as subgroups the transformation groups of practically all other geometries that had been studied. This is essentially what Cayley meant when he remarked that "projective geometry contains all geometry." Actually, as far as the theorems of the geometries are concerned, it is the other way about—the theorems of projective geometry are contained among the theorems of each of the other geometries.

Inasmuch as the product of any two homeomorphisms of the plane is a homeomorphism of the plane, and the inverse of any homeomorphism of the plane is also a homeomorphism of the plane, we see that the topology, or analysis situs, of the plane is a geometry in the Kleinian sense. And we also see that Cayley was not quite correct—that it is more proper to say that "topology contains all geometry." That is, all the planar transformation groups mentioned above are subgroups of the transformation group of all homeomorphisms of the plane. Plane topology emerges as the most fundamental of all the plane geometries— any theorem of plane topology holds within all the other plane geometries. Since the topological transformations are so broad, it is fair to wonder what properties of the plane of points can possibly remain invariant under these transformations. To give a few simple examples, we might mention, first of all, that a simple closed curve (a closed curve which does not cut itself) remains a simple closed curve under all topological transformations. Again, the fact that the deletion of only one point from a simple closed curve does not disconnect the curve is a topological property, and the fact that the deletion of two points from a simple closed curve separates the curve into two pieces is also a topological property

Some geometers prefer to modify slightly Klein's definition of a geometry to: A *geometry* is the study of those properties of a set S which remain invariant when the elements of S are subjected to the transformations of some transformation group Γ, *but which are not invariant under the transformations of any proper supergroup of* Γ. Under this modified definition, any theorem in a sequence of embracing geometries can belong to only one of the geometries of the sequence, and not to a certain geometry and also to all those preceding it in the sequence. This permits one to classify a theorem within an embracing sequence of geometries according to the geometry of the sequence to which it belongs.

For almost fifty years the Klein synthesis and codification of geometries remained essentially valid. But shortly after the turn of the century, bodies of mathematical propositions, which mathematicians felt should be called geometries, came to light; these bodies of propositions could not be fitted into this codification, and a new point of view upon the matter was developed, based upon the idea of abstract space with a superimposed structure that may or may not be definable in terms of some transformation group. Nevertheless, the Kleinian concept is still useful where it applies, and we might call a geometry that fits Klein's definition as given above a *Kleinian geometry*. Partially successful efforts have been made in the twentieth century, particularly by Oswald Veblen (1880–1960) and Élie Cartan (1869–1951), to extend and generalize Klein's definition so as to include geometries that lie outside Klein's original *Programm*.

[p] *Section 9.* We shall say that a boundary point in a planar map is common to n regions of the map if and only if an arbitrarily small circle about the point as center includes interior points of n of the regions. In Fig. 2 (page 3) we have a planar map of four regions in which each region is adjacent to the other three. For the most part, points on the boundaries between the regions are points common to just *two* of the regions. In fact, in the map in Fig. 2, there are only three points which are common to *three* of the regions, and no points common to all *four* of the regions.

Intuition tells us that points common to three or more regions of a planar map must be isolated points (as in the map in Fig. 2). In other words, intuition tells us that it is impossible for three or more regions to have a whole *line* of points in common.

Now our intuition is actually correct if, as we have been supposing in the past, the boundary of each region is a simple closed curve. But if we allow our regions to have more esoteric kinds of boundaries, it can be shown that our intuition is false. And this is precisely the aim of the early part of Section 9—to point out the great care that must be exercised when working in general set topology, since intuition can often lead us astray. As an example, before proceeding with the particular matter at hand, let us show that the famous four-color conjecture is false if we allow the regions of our map to possess boundaries which are not necessarily simple closed curves. In the map in Fig. 93, the wavy boundary between regions 4 and 5 is part of the graph of $y = \sin(1/x)$. Five colors are needed to color this map! But the two regions 4 and 5 do not have simple closed curves as boundaries.

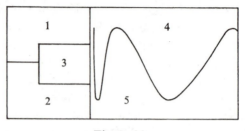

Figure 93

L. E. J. Brouwer (in *Mathematiche Annalen, 68* (1909), 427) showed that it is possible to have a planar map of three (four, five, $\cdots$, in fact, infinitely many) simply connected bounded regions[1] which all have the same boundary. We shall here content ourselves with showing that it is possible to have a planar map of three (four, five, $\cdots$) simply connected bounded regions which all have the same straight line segment as part

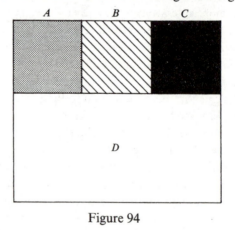

Figure 94

---

[1] A planar region is *connected* if any two of its points can be joined by a continuous curve belonging entirely to the region. The region is *simply connected* if it is connected and if every simple closed curve lying within the region can be shrunk to a point without leaving the region. The region is *bounded* if there exists a circle containing the region entirely within its interior.

of their boundaries. The construction we give is an adaptation, due to
H. Hahn, of Brouwer's original construction.

In Fig. 94 is pictured a map with three countries A, B, and C, together with a section D of unclaimed territory. Let us assume that D is
300 miles long and 150 miles wide. Country A commences by claiming
all the land in D which lies more than $50 = 150/3$ miles from the
boundary of D, plus a narrow corridor connecting the new territory to
the mother country (Fig. 95). Country B then claims all the remaining

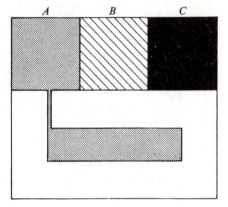

Figure 95

territory which lies more than $150/9$ miles from the new boundary of D,
plus a narrow corridor connecting the new territory to the mother
country (Fig. 96). Then country C annexes all the remaining territory

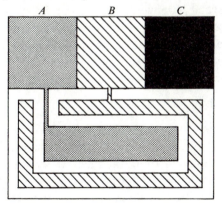

Figure 96

which lies more than 150/27 miles from the still newer boundary of D, plus a narrow corridor connecting the new territory to the mother country (Fig. 97). The countries then begin over again. A claims all the land which lies more than 150/81 miles from what is the new boundary of D, plus another narrow corridor connecting the newly annexed territory to the mother country; B claims all the land more than 150/243 miles from the new boundary, plus a corridor; C claims all the land

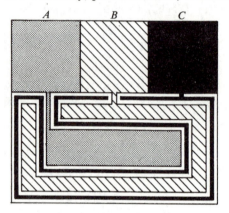

Figure 97

more than 150/729 miles from the still newer boundary, plus a corridor; and so on indefinitely.

In the limit, all of the original territory D will have been claimed. Moreover, this can be brought about in a finite length of time by assuming that the first annexation took place in half a year, the second in a quarter of a year, the third in an eighth of a year, and so on. For then the total time in years required to annex all the territory of D would be the sum of the infinite series

$$1/2 + 1/4 + 1/8 + \cdots + 1/2^n + \cdots,$$

which is one year.

Now consider any point P on, say, what was originally the lower boundary of D. Since an arbitrarily small circle with center P includes interior points of all three enlarged countries A, B, and C, we see that P is a boundary point of all three countries. It follows that all three countries have the line segment which was originally the lower boundary of D as a common boundary. Furthermore, it is clear that the same argument, given above for three countries, can be given for any number of countries. Needless to say, the boundaries of our countries are not simple closed curves.

While on the subject of intuition versus truth in set topology, let us consider one more problem. A sphere in space clearly enjoys the property that any circle in space exterior to the sphere can be shrunk,

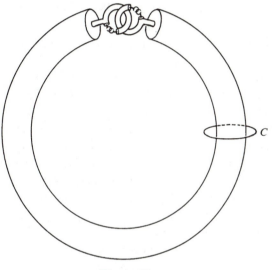

Figure 98

entirely in the exterior of the sphere, to a point. Can the same be said about every surface homeomorphic to a sphere? Intuition would seem to indicate an affirmative answer, but this is false, as was first shown by L. Antoine in 1921 (in [3]), and again later by J. W. Alexander in 1924 (*Proceedings of the National Academy of Science, 10* (1924), 6–12). Alexander's proof lies in the construction of the so-called *horned sphere*, illustrated in Fig. 98. A horned sphere may be described as follows. A long circular cylinder closed at both ends is bent until the two ends are near to each other. Then tubes are pushed out each end and hooked as shown. The process of pushing out additional tubes and hooking them is continued indefinitely. It can be shown that this horned sphere is homeomorphic to an ordinary sphere, but the circle C in Fig. 98, which

lies in space exterior to the horned sphere, cannot be shrunk, entirely in
the exterior of the horned sphere, to a point. The horned sphere thus
provides us with a counterexample to the once-believed conjecture that
if a surface S in space is homeomorphic to a sphere, then there is a
homeomorphism of the whole of space onto itself that carries S onto
the surface of a sphere. It is still unknown (1967) what restriction must
be imposed on S to insure that such a homeomorphism exists.

[q]   *Section 10.*   Topology, as a self-connected study, scarcely pre-
dates the mid-nineteenth century, but one can find some earlier isolated
topological investigations. Toward the end of the seventeenth century,
Leibniz used the term *geometria situs* to describe a sort of qualitative
mathematics that today would be thought of as topology, and he pre-
dicted important studies in this field, but his prediction was slow in
materializing. An early-discovered topological property of a simple
closed polyhedron is the relation $V - E + F = 2$, where V, E, and F
denote the number of vertices, edges, and faces, respectively, of the
polyhedron. This relation was known to Descartes in 1640, but the
first proof of the formula was given by Euler in 1752. Euler had earlier,
in 1736, considered some topology of linear graphs in his treatment of
the Königsberg bridge problem. Gauss made several contributions to
topology. Of the several proofs that he furnished of the fundamental
theorem of algebra, two are explicitly topological. His first proof of this
theorem employs topological techniques and was given in his doctoral
dissertation in 1799 when he was 22 years old. Later, Gauss briefly considered the theory of knots, which today is an important subject in
topology. About 1850 Francis Guthrie conjectured the four-color
problem, which was later taken up by Augustus De Morgan, Arthur
Cayley, and others. At this time the subject of topology was known as
analysis situs. The term *topology* was introduced by J. B. Listing, one of
Gauss's students, in 1847, in the title, *Vorstudien zur Topologie*, the first
book devoted to the subject. The German word *Topologie* was later
anglicized to *topology* by Professor Solomon Lefschetz of Princeton
University. G. B. Kirchoff, another of Gauss's students, in 1847 employed the topology of linear graphs in his study of electrical networks.
But of all of Gauss's students, the one who contributed by far the most
to topology was Bernhard Riemann, who, in his doctoral thesis of 1851,

introduced topological concepts into the study of complex-function theory. The chief stimulus to topology furnished by Riemann was his notion of *Riemann surface*, a topological device for rendering multiple-valued complex functions into single-valued functions. Also of importance in topology is Riemann's probationary lecture of 1854, concerning the hypotheses that lie at the foundations of geometry. This lecture furnished the breakthrough to higher dimensions, and the term and concept of *manifold* were introduced here. About 1865, A. F. Möbius wrote a paper in which a polyhedron was viewed simply as a collection of joined polygons. This introduced the concept of *2-complexes* into topology. In this systematic development of 2-complexes, Möbius was led to the surface now referred to as a *Möbius band* or a *Möbius strip*. In 1873, J. C. Maxwell used the topological theory of connectivity in his study of electromagnetic fields. Many other names, such as H. Helmholtz and Lord Kelvin, can be added to the list of physicists who applied topological ideas with success. Henri Poincaré ranks very high among the early contributors to topology. A paper of his, written in 1895 and entitled *Analysis situs*, is the first significant paper devoted wholly to topology. It was in that paper that the important homology theory of *n* dimensions was introduced. With Poincaré's work, the subject of topology was well under way, and an increasing number of mathematicians entered the field.

The notion of a geometric figure as made up of a finite set of joined fundamental pieces, as was emphasized by Möbius, Riemann, and Poincaré, gradually gave way to the Cantorian concept of an arbitrary set of points, and then it was recognized that any collection of things—be it a set of numbers, algebraic entities, functions, or nonmathematical objects—can constitute a topological space in some sense or other. This latter, and very general, viewpoint of topology has become known as *set topology*, whereas studies more intimately connected with the earlier viewpoint have become known as *combinatorial*, or *algebraic*, *topology*, although it must be confessed that this division is perhaps more one of convenience than of logic.

Topology started as a branch of geometry, but during the second quarter of the twentieth century it underwent such generalization and became involved with so many other branches of mathematics that it is now more properly considered, along with geometry, algebra, and analysis, as a fundamental division of mathematics. Today, topology may be roughly defined as the mathematical study of continuity.

[**r**]   *Section 11.*   The polygons are assumed to be simple closed polygons. Thus a cube with a similarly oriented smaller cube exteriorly attached to one of its faces is not a polyhedron, for in the resulting figure the concerned face has become ring-shaped and therefore is not a simple closed polygon. Of course, the resulting figure can be made into a polyhedron by dividing the ring-shaped face in two with, say, one of the diagonals of the outside square boundary of that face. This, in effect, increases the number of faces by one and the number of edges by two.

[**s**]   *Section 11.*   Diagrams of polyhedra of this sort are often called *Schlegel diagrams,* named after Victor Schlegel (1843–1905). If the

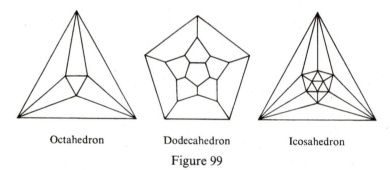

Octahedron   Dodecahedron   Icosahedron

Figure 99

polyhedron is convex, then a Schlegel diagram of the polyhedron can be obtained by a central projection of the polyhedron from an exterior point sufficiently close to the centroid of a selected face of the polyhedron; the projection of the selected face contains the projections of all the other faces. Fig. 14 shows Schlegel diagrams of the regular tetrahedron and the cube. In Fig. 99 are Schegel diagrams of the other regular polyhedra.

Two polyhedra that have the same configuration numbers n_v, n_e, and n_f are said to be *allomorphic*. That two allomorphic polyhedra need not be isomorphic (that is, have precisely the same configurational structure) is seen by the Schegel diagram in Fig. 100. This diagram represents a polyhedron which, like the cube, has six faces, twelve edges, and eight vertices, and which moreover, again like the cube, has three edges issuing from each vertex. But, unlike the cube, the faces of the polyhedron consist of two pentagons, two quadrilaterals, and two triangles.

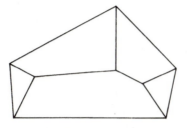

Figure 100

[t]   *Section 11.*   Many proofs have been given of Descartes' formula, $V - E + F = 2$, relating the number of vertices, edges, and faces of any simple closed polyhedron having simple closed polygons as faces. We here reproduce versions of a few famous attacks. We shall further assume, with no real loss of generality, that our polyhedra are convex.

I. EULER'S PROOF.   We shall construct the polyhedron by adding one face after another until the polyhedron is complete. (This actually reverses Euler's procedure, which was one of dissection instead of construction.) We assume it is always possible to add new faces in such a way that every new face has only *consecutive* edges in coincidence with edges of old faces. Set $V - E + F - 1 = \phi$.

For the first face, an n-gon, say, we have $V = n$, $E = n$, and $F = 1$, whence $\phi = 0$.

When a second face is added, say an m-gon having one edge in common with the n-gon, we have $V = n + m - 2$, $E = n + m - 1$, and $F = 2$, and again $\phi = 0$.

We now employ mathematical induction to prove that, as long as the polyhedron is incomplete, $\phi = 0$. Assume, then, that $\phi = 0$ at some stage of the construction when at least two faces of the polyhedron remain to be attached. In proceeding to the next stage, suppose a p-gon is added having q consecutive edges, and therefore $q + 1$ consecutive vertices, in coincidence with old faces. Letting unprimed letters refer to the previous stage and primed letters to the new stage, we then have

$$F' = F + 1, \quad E' = E + (p - q), \quad V' = V + (p - q - 1),$$

whence
$$\phi' = F' - E' + V' - 1 = F - E + V - 1 = \phi = 0.$$

It follows that, as long as the polyhedron is incomplete, $\phi = 0$.

Now, when the last face is added, no new edges or vertices are added, but the number of faces is increased by one. It follows that, for the completed polyhedron, $V - E + F - 1 = 1$, or $V - E + F = 2$.

II. LHUILIER AND STEINER'S PROOF.   Project the polyhedron orthogonally onto a plane, not perpendicular to any face of the polyhedron, to obtain a polygon P covered twice over with a network of subpolygons (see Fig. 101). This is always the case for a convex polyhedron, since no line can intersect such a polyhedron in more than two points.

Let S denote the sum of the angles of the subpolygons, and denote the number of sides of the F faces of the polyhedron by $n_1, n_2, \cdots, n_F$. Then

$$S = \sum_{i=1}^{F} (n_i - 2)\pi = \sum n_i\pi - 2F\pi = 2E\pi - 2F\pi = (E - F)2\pi.$$

Now if the bounding polygon P has n vertices, there are $V - n$ vertices interior to P, and we have

$$S = 2(n - 2)\pi + (V - n)2\pi = (V - 2)2\pi.$$

It follows that $E - F = V - 2$, or $V - E + F = 2$.

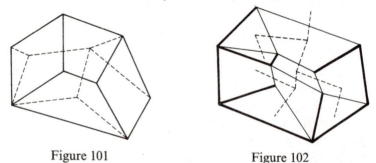

Figure 101   Figure 102

III. LEGENDRE'S PROOF.   Project the polyhedron from an interior point P onto the unit sphere with center P. Since every ray from P cuts the polyhedron exactly once, the sphere will be completely covered once over by a network of F spherical polygons. The area, on a unit sphere, of a spherical polygon of n sides and angle sum α, is $\alpha - (n - 2)\pi$.

Therefore if $n_1, n_2, \cdots, n_F$ denote the number of sides of the F spherical polygons, and if $\alpha_1, \alpha_2, \cdots, \alpha_F$ denote the angle sums for these polygons, the total area of the spherical polygons is

$$4\pi = \sum_{i=1}^{F} \alpha_i - \sum_{i=1}^{F} (n_i - 2)\pi$$

$$= \sum \alpha_i - \sum n_i\pi + 2F\pi$$

$$= 2V\pi - 2E\pi + 2F\pi,$$

whence $V - E + F = 2$.

IV. VON STAUDT'S PROOF. Consider a Schlegel diagram of the polyhedron (see Note [s]). Construct (see Fig. 102) a path starting from some selected vertex and leading along edges to every other vertex—this path may be branched anywhere, but it must not include any closed circuit. Next construct a similar kind of path starting within some selected face and, by crossing edges not used in the first path, leading to every other face. That this latter construction is possible follows from the fact that no face or group of faces is ring-fenced by edges of the first path. Also, there is only one path connecting two faces, for if there were two, this double path would isolate a group of vertices and show the first path to be incorrect. Thus, once the first path is chosen, the second path is uniquely determined and crosses each of the edges not employed in the first path.

Now the number of edges used in the first path is $V - 1$, and the number of edges crossed in the second path is $F - 1$. It follows that

$$(V - 1) + (F - 1) = E,$$

or $V - E + F = 2$.

[u] *Section 12.* One can deduce many interesting properties, which are by no means intuitively obvious, concerning simple closed polyhedra. For example, a tetrahedron and a pyramid with a square base are examples of simple closed polyhedra having precisely six and eight edges, respectively. It is natural to wonder if there is a simple closed polyhedron having exactly seven edges. We can prove there is no such polyhedron. We do this indirectly by assuming there is a simple closed polyhedron with exactly seven edges. Concentrate on any particular

face of the polyhedron and suppose that face has n sides. Since at least three edges issue from each vertex of this face, we see that $2n \leq 7$, or $n < 4$. It follows that all faces of the polyhedron must be triangles, whence $3F = 2E = 14$. But this is impossible, since F is an integer.

Many nonintuitive properties of simple closed polyhedra are consequences of the Descartes relation $V - E + F = 2$. Consider, for example, a simple closed polyhedron of V vertices, E edges, and F faces. Let F_n denote the number of faces having n sides, and let V_n denote the number of vertices from which n edges issue. Then

$$F = F_3 + F_4 + \cdots,$$
$$V = V_3 + V_4 + \cdots,$$
$$2E = 3F_3 + 4F_4 + 5F_5 + \cdots.$$

Now, using the relation $V - E + F = 2$, we have

$$2(F_3 + F_4 + \cdots) + 2(V_3 + V_4 + \cdots) = 4 + 3F_3 + 4F_4 + \cdots,$$

whence

(1) $2(V_3 + V_4 + \cdots) = 4 + F_3 + 2F_4 + 3F_5 + 4F_6 + \cdots$

and

(2) $2(F_3 + F_4 + \cdots) = 4 + V_3 + 2V_4 + 3V_5 + 4V_6 + \cdots.$

Doubling (2) and adding (1), we obtain

(3) $3F_3 + 2F_4 + F_5 = 12 + 2V_4 + 4V_5 + \cdots + F_7 + 2F_8 + \cdots.$

As some immediate consequences of (3) we have the following, where P stands for a simple closed polyhedron.

I. There is no P each of whose faces has more than five sides.

II. If P has no triangular or quadrilateral faces, then it has at least twelve pentagonal faces.

III. If P has no triangular or pentagonal faces, then it has at least six quadrilateral faces.

IV. If P has no quadrilateral or pentagonal faces, then it has at least four triangular faces.

We shall say P is *trihedral* if and only if precisely three edges issue from each vertex. Then:

V. If P is trihedral and has only pentagonal and hexagonal faces, the number of pentagonal faces is twelve.

VI. If P is trihedral and has only quadrilateral and hexagonal faces, the number of quadrilateral faces is six.

VII. If P is trihedral and has only triangular and hexagonal faces, the number of triangular faces is four.

Because of the symmetric role played by V and F in the formula $V - E + F = 2$, we also have the following *duals* of the above theorems.

I′. There is no P each of whose vertices has more than five edges issuing from it.

II′. If P has no trihedral or tetrahedral vertices, then it has at least twelve pentahedral vertices.

III′. If P has no trihedral or pentahedral vertices, then it has at least six tetrahedral vertices.

IV′. If P has no tetrahedral or pentahedral vertices, then it has at least four trihedral vertices.

V′. If P has only triangular faces and only pentahedral and hexahedral vertices, the number of pentahedral vertices is twelve.

VI′. If P has only triangular faces and only tetrahedral and hexahedral vertices, the number of tetrahedral vertices is six.

VII′. If P has only triangular faces and only trihedral and hexahedral vertices, the number of trihedral vertices is four.

Finally, by adding (1) and (2) above, we find

$$F_3 + V_3 = 8 + F_5 + V_5 + 2(F_6 + V_6) + \cdots,$$

whence it follows that, for any P, $F_3 + V_3 \geqq 8$. The tetrahedron, octahedron, and cube are examples of simple closed polyhedra for which $F_3 + V_3 = 8$.

[v]   *Section 14.*   A Möbius band is frequently called a *Möbius strip*.

[w]   *Section 14.*   A circular disk is a bilateral surface with one unknotted edge; a Möbius band is a unilateral surface with one unknotted edge. Do there exist, in ordinary three-dimensional space, bilateral and unilateral surfaces each having one *knotted* edge? The answer is

affirmative; Fig. 103 pictures the former (which was discovered by F. Frankl and L. S. Pontryagin in 1930), and Fig. 104 pictures the latter.

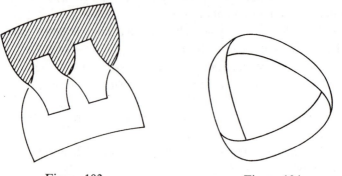

| Figure 103 | Figure 104 |

The interested reader might care to try to construct, in ordinary three-dimensional space, two-edged bilateral surfaces and two-edged unilateral surfaces where:

  a. the two edges are unknotted and unlinked,
  b. the two edges are unknotted and linked,
  c. the two edges are knotted and unlinked,
  d. the two edges are knotted and linked,
  e. one edge is unknotted and one knotted, and they are unlinked,
  f. one edge is unknotted and one knotted, and they are linked.

Solutions may be found in Martin Gardner, *The (First) Scientific American Book of Mathematical Puzzles and Diversions*, New York: Simon and Schuster, 1959, pages 63–72.

[x] *Section 15.* The Möbius band has been the object of "jokes," puzzles, and stories. For some of the stories, see, for example:

1. Nelson Bond, "The Geometrics of Johnny Day," *Astounding Science Fiction*, July, 1941.
2. W. H. Upson, "Alexander Botts and the Moebius Strip," *Saturday Evening Post*, Dec. 22, 1945.
3. Martin Gardner, "The No-sided Professor," *Esquire*, Jan., 1947.
4. W. H. Upson, "Paul Bunyan vs. the Conveyor Belt," *Ford Times*, July, 1949.

5. A. J. Deutsch, "A Subway Named Moebius," *Street and Smith*, 1950. (Stories 2, 3, and 5 have been reprinted in Clifton Fadiman, *Fantasia Mathematica*, New York: Simon and Schuster, 1958.)

Among the puzzles connected with the Möbius band are those based on cutting the band along the median line and then the result along its median line, or by cutting the Möbius band along the line one-third the width of the band from the edge of the band, etc. For some of these recreational puzzles, see W. W. Rouse Ball and H. S. M. Coxeter, *Mathematical Recreations and Essays*, New York: The Macmillan Company, 1939, pages 125–126.

Of the "jokes" we might mention the following:

1. Consider the plight of two painters on opposite faces of a Möbius band, one painting with white paint and the other with black paint.

2. Consider the wisdom in the procedure of the witch doctor who gave advice to couples wondering if they should get married. If he wished to prophesy a future break-up in the proposed marriage, he would split an untwisted band; if he wished to prophesy that the couple would quarrel but still stay together, he would split a band having a full twist; if he wished to prophesy a perfect marriage, he would split a Möbius band.

3.   A mathematician confided
    That a Möbius strip is one-sided,
      And you'll get quite a laugh
      If you cut one in half,
    For it stays in one piece when divided.

4.   A burleycue dancer, a pip
    Named Virginia, could peel in a zip,
      But she read science fiction
      And died of constriction
    Attempting a Moebius strip.

[y]   *Section 15.*   One is reminded of the proof by "logic," in *Robin Hood* (Chapter 14, "Robin Meets Friar Tuck"), that *every* river has only one bank:

"Well then, good fellow, holy father, or whatever you are," said Robin, "I would like to know if this same friar lives on this side of the river or the other."

"Truly, the river has no side but the other," said the friar.

"How can you prove that?" asked Robin.

"Why, this way," said the friar, noting the points on his fingers. "The other side of the river is the other, right?"

"Yes, that's true."

"Yet the other side is but one side, do you see?"

"Nobody could say it isn't," said Robin.

"Then if the other side is one side, this side is the other side. But the other side is the other side; therefore both sides of the river are the other side, just as I said."

"That is a good argument," said Robin. "Yet I still do not know whether the friar I am seeking is on the side of the river where we are standing or on the side where we are not."

"That," said the friar, "is a question which the rules of argument can't answer. I advise you to find out by the use of your senses such as sight, feeling, and such things."

[z] *Section 17.* A generalized torus with p holes is frequently called a *sphere with p handles*, inasmuch as a generalized torus with p holes is homeomorphic to the surface of a sphere to which p teacup handles have been attached. The ordinary torus is thus a sphere with one handle.

[aa] *Section 18.* A Klein surface is frequently referred to as a *Klein bottle*.

[bb] *Section 18.* One recalls the limerick:

> A mathematician named Klein
> Thought the Möbius strip was divine.
>   Said he, "If you glue
>   The edges of two
> You'll get a weird bottle like mine."

[cc] *Section 25.* The proof, in Sections 24 and 25, of the classification theorem is based upon the proof given by H. R. Brahana, "Systems of circuits on two-dimensional manifolds," *Annals of Mathematics, 23* (1921–1923), 144–168. Brahana's procedure has been reproduced, with modifications, in several places, such as DE KERÉKJÁRTÓ [14], Chapter 2,

Sections 6 and 7; SEIFERT and THRELFALL [22], Sections 37–39; S. S. Cairns, *Introductory Topology*, New York: The Ronald Press Company, 1961, Sections 2-4 and 2-5; Howard Eves, *A Survey of Geometry*, Vol. 2, Boston: Allyn and Bacon, Inc., 1965, Section 15.3. An alternative approach may be found in R. C. James, "Combinatorial topology of surfaces," *Mathematics Magazine*, 29, No. 1 (Sept.–Oct., 1955), 1–39. This paper is based on lectures given by Professor A. W. Tucker while Philips lecturer at Haverford College in the fall semester of 1953–1954; a summary of the method can be found in Howard Eves, *loc. cit.*, Exercises 15.3-5 and 15.3-6.

[**dd**] *Section 28.* Consider two concentric spherical surfaces S_1 and S_2, S_1 inside S_2. Let points of S_1 and S_2 on the same radial ray be called a pair of associated points. Let A_1, A_2 and B_1, B_2 be two pairs of associated points, and let c_1 joining A_1 and B_1 on S_1 and c_2 joining A_2 and B_2 on S_2 be a pair of associated simple arcs. Slit the surfaces along these curves c_1 and c_2. Denote the lips of cut c_1 by c_1^+ and c_1^- and the associated lips of cut c_2 by c_2^+ and c_2^-. Now join S_1 and S_2 by pasting together c_1^+ and c_2^- and also c_1^- and c_2^+. Consider the pairs of points A_1, A_2 and B_1, B_2 as identified, and the pair of curves c_1 and c_2 as a curve of self-penetration on the resulting "woven" surface formed by the two joined spheres. Whenever, in traveling on the woven surface, we cross the curve of self-penetration, we move on a ramp from one of the spherical surfaces to the other.

Now imagine n concentric spherical surfaces woven together in pairs along as many nonintersecting simple arcs for each pair as desired. The resulting woven surface formed from the n spherical surfaces is known as a *Riemann surface*, the spherical surfaces are known as the *sheets* of the Riemann surface, the curves of self-penetration are known as *cut lines*, and the endpoints of the cut lines are known as *branch points*.

As a simple example, consider a Riemann surface formed of two sheets S_1 and S_2 woven together along two cut lines c and d. Slit each sheet along c and open these slits, denoting the lips of the cut along c on S_1 by c_1^+ and c_1^- and those on S_2 by c_2^+ and c_2^-. Similarly treat the cut line d. We now have two detached concentric spherical surfaces with two holes cut in each surface. By a continuous deformation, we can slip the inner surface S_1 through one of the holes in the outer surface S_2, giving us the two spherical surfaces pictured in Fig. 105. By continuous deformation, these two open surfaces may be converted into those

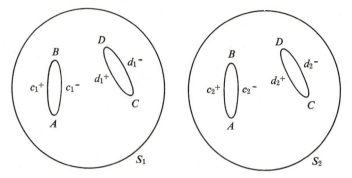

Figure 105

represented in Fig. 106. Finally, the two open surfaces of Fig. 106 may be rejoined, by pasting c_1^+ to c_2^-, c_1^- to c_2^+, d_1^+ to d_2^-, d_1^- to d_2^+, to form a torus (Fig. 107). It follows that our particular Riemann surface is homeomorphic to a torus, or a sphere with one handle.

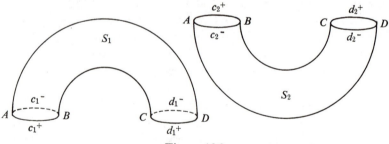

Figure 106

The woven surface in the above example is the Riemann surface that uniformizes the complex function

$$w^2 = (1 - z^2)(1 - k^2 z^2).$$

Each complex value of z, except $z = \pm 1$ and $z = \pm 1/k$, determines two values of w. In order to convert this function into a single-valued continuous function of z, one introduces two complex z-planes, I and II, one just above the other. These two planes are then woven together by ramps along joins connecting the points $z = 1$ and $z = 1/k$ and the

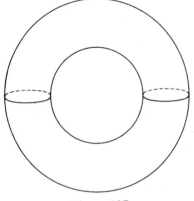

Figure 107

points $z = -1$ and $z = -1/k$, as schematically pictured in Fig. 108. Now this woven two-sheeted complex z-plane can be mapped by stereographic projection onto the woven two-sheeted surface of our example.

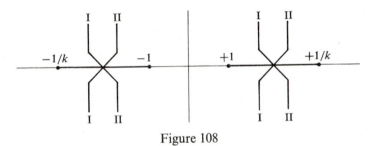

Figure 108

[ee] *Bibliography.* Both the textbook and the periodical literature bearing on topology have grown enormously in recent years, and this literature is written on all levels of difficulty. Following is a partial list of some of this literature; the reader should note that most of the items date from 1950—indeed, from 1960. One can expect this literature to reach deluge proportions during the next couple of decades. The following list is largely restricted to textbooks, and almost all are in English. The journal literature is, for the most part, only for the expert.

ALEKSANDROV, P. S., *Combinatorial Topology*, Vol. 1, tr. by H. Komm. Rochester, N.Y.: Graylock Press, 1956.

ALEXANDROFF, PAUL, *Elementary Concepts of Topology*, tr. by A. E. Farley. New York: Dover Publications, Inc., 1961.

ARNOLD, B. H., "A topological proof of the fundamental theorem of algebra," *American Mathematical Monthly, 56* (1949), 465–466.

ARNOLD, B. H., *Intuitive Concepts of Topology*. Englewood Cliffs, N.J.: Prentice-Hall, Inc., 1962.

AYRES, W. L., "Some elementary aspects of topology," *American Mathematical Monthly, 45* (1938), 88–92.

BALL, W. W. R., *Mathematical Recreations and Essays,* 11th ed., rev. by H. S. M. Coxeter. New York: Dover Publications, Inc., 1987.

BAUM, J. D., *Elements of Point Set Topology.* New York: Dover Publications, Inc., 1991.

BERGE, CLAUDE, *The Theory of Graphs and Its Applications.* Mineola, New York: Dover Publications, Inc., 2001.

BERGE, CLAUDE, *Topological Spaces*, tr. by E. M. Patterson. New York: The Macmillan Company, 1963.

BING, R. H., "Examples and counterexamples," *Pi Mu Epsilon Journal, 1* (1953), 311–317.

BING, R. H., "Point set topology," Chap. 10 in *Insights into Modern Mathematics*, Twenty-third Yearbook of the National Council of Teachers of Mathematics. Washington, D.C.: National Council of Teachers of Mathematics, 1957.

BING, R. H., *Elementary Point Set Topology*, Herbert Ellsworth Slaught Memorial Paper No. 8. Buffalo, N.Y.: Mathematical Association of America, 1960.

BOURGIN, D. G., *Modern Algebraic Topology.* New York: The Macmillan Company, 1963.

BRAHANA, H. R., "Systems of circuits on two-dimensional manifolds," *Annals of Mathematics, 23* (1921–1923), 144–168.

BRAHANA, H. R., "The four-color problem," *American Mathematical Monthly, 30* (1923), 234–243.

BUSACKER, R. G., and T. L. SAATY, *Finite Graphs and Networks: An Introduction with Applications.* New York: McGraw-Hill Book Company, 1965.

BUSHAW, D., *Elements of General Topology.* New York: John Wiley & Sons, Inc., 1963.

CAIRNS, S. S., *Introductory Topology*. New York: The Ronald Press Company, 1961.

CHINN, W. G., and N. E. STEENROD, *First Concepts of Topology*, New Mathematical Library, Monograph No. 18. New York: Random House, 1966.

COURANT, RICHARD, and HERBERT ROBBINS, *What Is Mathematics?* New York: Oxford University Press, 1941.

COXETER, H. S. M., "The map-coloring of unorientable surfaces," *Duke Mathematical Journal, 10* (1943), 293–304.

COXETER, H. S. M., *Regular Polytopes*. New York: Pitman Publishing Corporation, 1947.

COXETER, H. S. M., "The four-color map problem, 1840–1890," *The Mathematics Teacher, 52* (1959), 283–289.

COXETER, H. S. M., *Introduction to Geometry*. New York: John Wiley & Sons, Inc., 1961.

CROWELL, R. H., and R. H. FOX, *Introduction to Knot Theory*. Boston: Ginn & Company, 1963.

DELACHET, ANDRÉ, *Contemporary Geometry*, tr. by H. G. Bergmann. New York: Dover Publications, Inc., 1962.

DUGUNDJI, JAMES, *Topology*. Boston: Allyn and Bacon, Inc., 1966.

EVES, HOWARD, *A Survey of Geometry*, Vol. 2. Boston: Allyn and Bacon, Inc., 1965.

FADIMAN, CLIFTON, *Fantasia mathematica* (a set of stories and oddments drawn from mathematics, assembled and edited by Clifton Fadiman). New York: Simon and Schuster, Inc., 1958.

FRANKLIN, PHILIP, "The four color problem," in *Galois Lectures*, Scripta Mathematica Library Number Five. New York: Yeshiva University, 1941.

FRÉCHET, MAURICE, "Abstract sets, abstract spaces and general analysis," Chap. 20 in *The Tree of Mathematics* (ed. by Glenn James). Pacoima, Calif.: The Digest Press, 1957.

FRENCHEL, W. F., "Elementaire Beweise und Anwendungen einiger Fixpunktsätze," *Matematisk Tidsskrift, B* (1932), 66–87.

GAAL, S. A., *Point Set Topology*. New York: Academic Press, 1964.

GARDNER, MARTIN, "Topology and magic," *Scripta Mathematica, 17* (1951), 75–83.

GARDNER, MARTIN, "Topology: a strange new mathematics," *Science World, 1* (1957), 7–9.

GARDNER, MARTIN, *The Scientific American Book of Mathematical Puzzles and Diversions.* New York: Simon and Schuster, 1959.

GARDNER, MARTIN, *The 2nd Scientific American Book of Mathematical Puzzles and Diversions.* New York: Simon and Schuster, 1961.

GOLOVINA, L. I., and I. M. YAGLOM, *Induction in Geometry,* tr. by A. W. Goodman and Olga A. Titelbaum. Boston: D. C. Heath and Company, 1963.

HAHN, HANS, "Geometry and intuition," *Scientific American, 190* (1954), 84–91.

HALL, D. W., and G. L. SPENCER II, *Elementary Topology.* New York: John Wiley & Sons, Inc., 1955.

HALL, D. W., "Some concepts of elementary topology," Chap. 15 in *The Tree of Mathematics* (ed. by Glenn James). Pacoima, Calif.: The Digest Press, 1957.

HILBERT, DAVID, and S. COHN-VOSSEN, *Geometry and the Imagination,* tr. by P. Nemenyi. New York: Chelsea Publishing Company, 1952.

HOCKING, J. G., "Topology," *Astounding Science Fiction, 53* (1954), 96–110.

HOCKING, J. G., and G. S. YOUNG, *Topology.* New York: Dover Publications, Inc., 1987.

HU, SZE-TSEN, *Elements of General Topology.* San Francisco: Holden-Day, Inc., 1964.

HUREWICZ, WITOLD, and HENRY WALLMAN, *Dimension Theory.* Princeton, N.J.: Princeton University Press, 1948.

JAMES, R. C., "Combinatorial topology of surfaces," *Mathematics Magazine, 29* (1955), 1–39.

JONES, B. M., *Elementary Concepts of Mathematics,* 2nd ed. New York: The Macmillan Company, 1963.

KASNER, EDWARD, and JAMES NEWMAN, *Mathematics and the Imagination.* New York: Simon and Schuster, 1940.

KELLEY, J. L., *General Topology.* Princeton, N.J.: D. Van Nostrand Company, Inc., 1955.

KLINE, J. R., "What is the Jordan curve theorem?" *American Mathematical Monthly, 49* (1942), 281–286.

KOWALSKY, H. J., *Topological Spaces*, tr. by J. E. Strum. New York: Academic Press, 1964.

KURATOWSKI, KAZIMIERZ, "Sur le problème des courbes gauches en topologie," *Fundamentae Mathematicae, 15* (1930), 271–283.

KURATOWSKI, KAZIMIERZ, *Introduction to Set Theory and Topology*, tr. by L. F. Boron. New York: Pergamon Press, Inc., 1961.

LEFSCHETZ, SOLOMON, *Introduction to Topology*. Princeton, N.J.: Princeton University Press, 1949.

LEFSCHETZ, SOLOMON, *Topology*, 2nd ed. New York: Chelsea Publishing Company, 1956.

LIETZMANN, WALTHER, *Anschauliche Topologie*. Munich: R. Oldenbourg, 1955.

MAMUZIĆ, Z. P., *Introduction to General Topology*, tr. by L. F. Boron. Groningen, The Netherlands: P. Noordhoff, Ltd., 1963.

MANHEIM, J. H., *The Genesis of Point Set Topology*. New York: The Macmillan Company, 1964.

MANSFIELD, MAYNARD, *Introduction to Topology*. Princeton, N.J.: D. Van Nostrand Company, Inc., 1963.

MASSEY, W. S., "Topology, algebraic," *The Encyclopaedia Britannica*, Vol. XXII, 1964.

MENDELSON, BERT, *Introduction to Topology*. New York: Dover Publications, Inc., 1990.

MENGER, KARL, "What is dimension?" *American Mathematical Monthly, 50* (1943), 2–7.

MESERVE, B. E., "Topology for secondary schools," *The Mathematics Teacher, 46* (1953), 465–474.

MESERVE, B. E., *Fundamental Concepts of Geometry*. New York: Dover Publications, Inc., 1983.

MONTGOMERY, DEANE, "What is a topological group?" *American Mathematical Monthly, 52* (1945), 302–307.

MOORE, R. L., *Foundations of Point Set Topology*, American Mathematical Society Colloquium Publications, Vol. XIII, rev. ed. Providence, R.I.: American Mathematical Society, 1962.

NAGATA, JUN-ITI, *Modern Dimension Theory*. Amsterdam, The Netherlands: North Holland Publishing Company, 1965.

NEWMAN, M. H. A., *Elements of the Topology of Plane Sets of Points*, 2nd ed. New York: Cambridge University Press, 1954.

OGILVY, C. S., *Through the Mathescope* (Chap. 11). New York: Oxford University Press, 1956.

OGILVY, C. S., *Tomorrow's Math* (Chap. 6). New York: Oxford University Press, 1962.

ORE, OYSTEIN, *Graphs and Their Uses*, New Mathematical Library, Monograph No. 10. New York: Random House, 1963.

ORE, OYSTEIN, *Theory of Graphs*, American Mathematical Society Colloquium Publications, Vol. XXXVIII. Providence, R.I.: American Mathematical Society, 1962.

PATTERSON, E. M., *Topology*. New York: Interscience Publishers, Inc., 1956.

PERVIN, W. J., *Foundations of General Topology*. New York: Academic Press, 1964.

PONTRYAGIN, L. S., *Foundations of Combinatorial Topology*, tr. by F. Bagemihl, H. Komm, and W. Seidel. Rochester, N.Y.: Graylock Press, 1952.

RADEMACHER, HANS and OTTO TOEPLITZ, *The Enjoyment of Mathematics*. New York: Dover Publications, Inc., 1990.

RADÓ, TIBOR, "What is the area of a surface?" *American Mathematical Monthly*, 50 (1943), 139–141.

SIMMONS, G. F., *Introduction to Topology and Modern Analysis*. New York: McGraw-Hill Book Company, Inc., 1963.

SOMMERVILLE, D. M. Y., *An Introduction to the Geometry of n Dimensions*. New York: Dover Publications, Inc., 1958.

STEIN, S. K., *Mathematics, the Man-made Universe*. Mineola, New York: Dover Publications, Inc., 1999.

THRON, W. J., *Topological Structures*. New York: Holt, Rinehart and Winston, 1966.

TUCKER, A. W., "Some topological properties of disk and sphere," *Proceedings of the First Canadian Mathematical Congress*. Toronto: The University of Toronto Press, 1946, 285–309.

TUCKER, A. W., and H. S. BAILEY, JR., "Topology," *Scientific American*, 182 (1950), 18–24.

TUCKERMAN, B. A., "A nonsingular polyhedral Möbius band whose boundary is a triangle," *American Mathematical Monthly*, 55 (1948), 309–311.

VAIDYANATHASWAMY, R., *Set Topology,* 2nd ed. Mineola, New York: Dover Publications, Inc., 1999.

WALLACE, A. H., *An Introduction to Algebraic Topology.* New York: Pergamon Press, 1957.

WHYBURN, G. T., "What is a curve?" *American Mathematical Monthly,* *49* (1942), 493–497.

WILDER, R. L., "Some unsolved problems in topology," *American Mathematical Monthly,* *44* (1937), 61–70.

WILDER, R. L., *Topology of Manifolds,* American Mathematical Society Colloquium Publications, Vol. XXXII. Providence, R.I.: American Mathematical Society, 1949.

WILDER, R. L., "Topology, general," *The Encyclopaedia Britannica,* Vol. XXII, 1964.

Index

Abstract topology, 19
Alexander, J. W., 19, 99
Alexandroff, P., 19
Algebraic topology, 101
Allomorphic, 102
Analysis situs, 100
Annulus, 75
Antoine, L., 99
Ascoli, G., 19

Betti, E., 19
Bicontinuous transformation, 8
Bilateral surface, 29
Birkhoff, George, 20
Biunique correspondence, 8
Biunique transformation(s), 8, 90
   identical, 91
   inverse of, 91
   product of, 90
Boundary point, 95
Bounded planar region, 96 (*foot-note*)
Brouwer, L. E. J., 19, 96

Cantor, G., 17, 19
Cartan, É., 95
Cayley, A., 2, 79, 94, 100
Characteristic of a polyhedron, 65
Characteristic of a surface, 27–29
Chromatic number, 3
Circle geometry, 93
Closed Jordan curve, 2, 13
Closed polyhedral surface, 45

Closed surface, 46
Combinatorial topology, 18, 101
Complex, 18
Connected planar region, 96 (*foot-note*)
Connection number of a surface, 71
Continuous transformation, 8
Coxeter, H. S. M., 83
Cross cap, 41
Curve with one bank, 32
Curvilinear polygon, 36

Dehn, M., 67
De Morgan, A., 2, 79, 100
Descartes, R., 100
Descartes' formula, 21
   alternative proofs of, 103–105
   some consequences of, 106, 107
Descartes' theorem, 21*ff*, 22
   applications of, 25*ff*
Disjoint sets, 12 (*footnote*)
Dyadic relation, 89
   reflexive, 89
   symmetric, 89
   transitive, 89
Dyck, W. von, 19

Elementarily associated polyhe-dra, 55
Elementary operations, 53
Equivalence relation, 89
Equivalent sides and vertices, 50

Erlanger Programm, 90
Euclid, vi
Euler, L., 19, 21, 100
Euler's proof of Descartes' formula, 103
Exterior of simple closed polygon, 78

Fan, K., ix, x
Figures with same position in space, 15
Five-color theorem, 81
Four-color conjecture, 2
  history of, 78–80
Franklin, P., 79, 80
Fréchet, M., ix, x, 19

Gauss, K. F., 15, 19, 100
Generalized torus with p holes, 39
General topology, 19
Genus of a surface, 70
Geometria situs, 100
Geometry
  circle, 93
  India-rubber, 6
  Kleinian, 95
  Klein's definition of, 92, 94
  line, 93
  pelastic-sheet, 88
  plane affine, 93
  plane centro-affine, 93
  plane equiform, 92
  plane Euclidean metric, 92
  plane projective, 92
  plane similarity, 92
  point, 93
  rubber-sheet, 88
  sphere, 93
Group, 90
  transformation, 92

Guthrie, Francis, 2, 78, 100
Guthrie, Frederick, 2, 4, 79

Hadamard, J., 19
Hahn, H., 97
Heawood, P. J., 79, 82, 83, 84
Heawood's formula, 83
Heawood's theorem, 79, 84
Heegaard, P., 67
Heffter, L., 83
Helmholtz, H., 101
Heptahedron of C. Reinhardt, 35
Homeomorphic figures, 10
Homeomorphism, 8
Homotopic figures, 17
Homotopic sets in a set, 17
Homotopy, 16
Homotopy on a set, 17
Hopf, H., 19
Hoppe curve, 14 (footnote)
Horned sphere, 99

Identical transformation, 91
Identified sides and vertices, 50
Image of a point, 8
India-rubber geometry, 6
Interior of simple closed polygon, 78
Inverse transformation, 8, 91
Isotopic figures, 16
Isotopic sets in a set, 17
Isotopy, 9, 16
Isotopy on a set, 17

Jordan, C., 19, 76
Jordan curve theorem, 2, 75–78

Kagno, I. N., 83
Kellog, O. D., 20
Kelvin, Lord, 101

Kempe, A. B., 79
Kirchoff, G. B., 100
Klein, F., 90
Klein bottle, 110
Kleinian geometry, 95
Klein's definition of a geometry, 92, 94
Klein surface, 41
    limerick, 110
Kronecker, L., 19
Kuratowski, K., 88

Lebesgue, H. L., 19
Leech, J., 84
Lefschetz, S., 19, 20, 100
Legendre's proof of Descartes' formula, 104
Leibniz, G. W., 19, 100
Lhuilier's proof of Descartes' formula, 104
Lie, S., 90
Line geometry, 93
Linking number, 15
Listing, J. B., 29 (*footnote*), 100
Lyusternik, L. A., 20

Manifold, 101
Maximum number of regions adjacent in pairs, 5
Maxwell, J. C., 101
Möbius, A. F., 19, 29 (*footnote*), 78, 101
Möbius band, 29, 101
    limericks, 109
Möbius's rule of edges, 34
Möbius strip, 101, 107
Morse, M., 20

Nonorientable polyhedron, 65
Nonorientable surface, 31

Nonorientable torus, 41
Normal form of a polyhedron, 55

One-dimensional subdivision, 53
One-dimensional union, 54
One-to-one correspondence, 88
Orientable polyhedron, 65
Orientable surface, 31

Pelastic-sheet geometry, 88
Planar polygonal schema, 50, 51
Plane affine geometry, 93
Plane centro-affine geometry, 93
Plane equiform geometry, 92
Plane Euclidean metric geometry, 92
Plane projective geometry, 92
Plane similarity geometry, 92
Plato, 27
Platonic solids, 27
Poincaré, H., v, 19, 20, 21, 101
Point geometry, 93
Point transformation, 8 (*footnote*)
Polygon
    curvilinear, 36
    side of, 36
    topological, 36
    vertex of, 36
Polygonal division, 46
Polyhedron, polyhedra, 21, 45
    characteristic of, 65
    edge of, 21, 46
    elementarily associated, 55
    face of, 21, 46
    nonorientable, 65
    normal form of, 55
    orientable, 65
    simple, 22
    symbolic representation of, 51
    topological, 45

Polyhedron, polyhedra—*cont.*
trihedral, 106
vertex of, 46
Poncelet, V., 12, 13
Principal problem in the topology of closed surfaces, 49, 68
Problem of coloring geographic maps, 2
Problem of neighboring regions, 5
Projective figures, 13

Reduction to normal form, 56–64
Reflexive dyadic relation, 89
Regular polyhedra, 25–27
Reinhardt, C., 35
heptahedron of, 35
Relative topological property, 12, 14*ff*, 15
Reynolds, C. N., 80
Riemann, G. F. B., vi, 19, 20, 100
Riemann surface, 69, 101, 111–113
branch point of, 111
cut line of, 111
sheets of, 111
Ringel, G., 84
Rubber-sheet geometry, 88

Schauder, J., 20
Schläfli, L., 19
Schlegel, V., 102
Schlegel diagram, 102
Schnirelmann, L., 20
Set topology, 17, 101
Seven-color theorem, 79
Severi, F., 20
Simple polyhedron, 22
Simply connected planar region, 96 (*footnote*)
Sphere geometry, 93

Sphere with one cross cap, 43
Sphere with *p* handles, 110
Sphere with *q* crosscaps, 44
Sphere with two cross caps, 43
Steiner's calyx surface, 35 (*footnote*)
Steiner's proof of Descartes' formula, 104
Steiner's Roman surface, 35 (*footnote*)
Stereographic projection, 10
Symbolic representation of a polyhedron, 51
Symmetric dyadic relation, 89

Tait, P. G., 79
Tetrahedral division, 28
Theorem
five-color, 81
Heawood's, 79, 84
Jordan curve, 75–78
of M. Dehn and P. Heegaard, 67
principal, of the topology of closed surfaces, 68
seven-color, 79
Thread problem, 87
"Three houses and three wells" puzzle, 87
Topological classes, 12
Topological equivalence, 12
Topological invariant, 11
Topological polygon, 36
Topological polyhedron, 45
Topological property, 7, 11
relative, 11
Topological transformation, 8
Topology, 7, 11
algebraic, 101
combinatorial, 18, 101

Topology—*cont.*
  development (or history) of, 19–
    20, 100–102
  origin of term, 100
  set, 17, 101
Torus, 3
Transformation, 8
  bicontinuous, 8
  biunique, 8, 90
  continuous, 8
  group, 92
  identical, 91
  inverse, 8, 91
  point, 8 (*footnote*)
  topological, 8
Transform of a point, 8
Transitive dyadic relation, 89

Trihedral polyhedron, 106
Two-complex, 101
Two-dimensional subdivision, 54
Two-dimensional union, 54

Unilateral surface, 30

Vandermonde, A. T., 19
Van der Waerden, B. L., 20
Veblen, O., 19, 95
Volterra, V., 19
Von Staudt's proof of Descartes'
  formula, 105

Weiske, 78
Winn, C. E., 80

A CATALOG OF SELECTED
DOVER BOOKS
IN SCIENCE AND MATHEMATICS

A CATALOG OF SELECTED

DOVER BOOKS

IN SCIENCE AND MATHEMATICS

Astronomy

BURNHAM'S CELESTIAL HANDBOOK, Robert Burnham, Jr. Thorough guide to the stars beyond our solar system. Exhaustive treatment. Alphabetical by constellation: Andromeda to Cetus in Vol. 1; Chamaeleon to Orion in Vol. 2; and Pavo to Vulpecula in Vol. 3. Hundreds of illustrations. Index in Vol. 3. 2,000pp. 6⅛ x 9¼.
23567-X, 23568-8, 23673-0 Three-vol. set

THE EXTRATERRESTRIAL LIFE DEBATE, 1750–1900, Michael J. Crowe. First detailed, scholarly study in English of the many ideas that developed from 1750 to 1900 regarding the existence of intelligent extraterrestrial life. Examines ideas of Kant, Herschel, Voltaire, Percival Lowell, many other scientists and thinkers. 16 illustrations. 704pp. 5⅜ x 8½.
40675-X

A HISTORY OF ASTRONOMY, A. Pannekoek. Well-balanced, carefully reasoned study covers such topics as Ptolemaic theory, work of Copernicus, Kepler, Newton, Eddington's work on stars, much more. Illustrated. References. 521pp. 5⅜ x 8½.
65994-1

AMATEUR ASTRONOMER'S HANDBOOK, J. B. Sidgwick. Timeless, comprehensive coverage of telescopes, mirrors, lenses, mountings, telescope drives, micrometers, spectroscopes, more. 189 illustrations. 576pp. 5⅝ x 8¼. (Available in U.S. only.)
24034-7

STARS AND RELATIVITY, Ya. B. Zel'dovich and I. D. Novikov. Vol. 1 of *Relativistic Astrophysics* by famed Russian scientists. General relativity, properties of matter under astrophysical conditions, stars, and stellar systems. Deep physical insights, clear presentation. 1971 edition. References. 544pp. 5⅝ x 8¼. 69424-0

Chemistry

CHEMICAL MAGIC, Leonard A. Ford. Second Edition, Revised by E. Winston Grundmeier. Over 100 unusual stunts demonstrating cold fire, dust explosions, much more. Text explains scientific principles and stresses safety precautions. 128pp. 5⅜ x 8½.
67628-5

THE DEVELOPMENT OF MODERN CHEMISTRY, Aaron J. Ihde. Authoritative history of chemistry from ancient Greek theory to 20th-century innovation. Covers major chemists and their discoveries. 209 illustrations. 14 tables. Bibliographies. Indices. Appendices. 851pp. 5⅜ x 8½. 64235-6

CATALYSIS IN CHEMISTRY AND ENZYMOLOGY, William P. Jencks. Exceptionally clear coverage of mechanisms for catalysis, forces in aqueous solution, carbonyl- and acyl-group reactions, practical kinetics, more. 864pp. 5⅜ x 8½.
65460-5

THE HISTORICAL BACKGROUND OF CHEMISTRY, Henry M. Leicester. Evolution of ideas, not individual biography. Concentrates on formulation of a coherent set of chemical laws. 260pp. 5⅜ x 8½.   61053-5

A SHORT HISTORY OF CHEMISTRY, J. R. Partington. Classic exposition explores origins of chemistry, alchemy, early medical chemistry, nature of atmosphere, theory of valency, laws and structure of atomic theory, much more. 428pp. 5⅜ x 8½. (Available in U.S. only.)   65977-1

GENERAL CHEMISTRY, Linus Pauling. Revised 3rd edition of classic first-year text by Nobel laureate. Atomic and molecular structure, quantum mechanics, statistical mechanics, thermodynamics correlated with descriptive chemistry. Problems. 992pp. 5⅜ x 8½.   65622-5

Engineering

DE RE METALLICA, Georgius Agricola. The famous Hoover translation of greatest treatise on technological chemistry, engineering, geology, mining of early modern times (1556). All 289 original woodcuts. 638pp. 6¾ x 11.   60006-8

FUNDAMENTALS OF ASTRODYNAMICS, Roger Bate et al. Modern approach developed by U.S. Air Force Academy. Designed as a first course. Problems, exercises. Numerous illustrations. 455pp. 5⅜ x 8½.   60061-0

DYNAMICS OF FLUIDS IN POROUS MEDIA, Jacob Bear. For advanced students of ground water hydrology, soil mechanics and physics, drainage and irrigation engineering and more. 335 illustrations. Exercises, with answers. 784pp. 6⅛ x 9¼.
                                                                          65675-6

ANALYTICAL MECHANICS OF GEARS, Earle Buckingham. Indispensable reference for modern gear manufacture covers conjugate gear-tooth action, gear-tooth profiles of various gears, many other topics. 263 figures. 102 tables. 546pp. 5⅜ x 8½.
                                                                          65712-4

MECHANICS, J. P. Den Hartog. A classic introductory text or refresher. Hundreds of applications and design problems illuminate fundamentals of trusses, loaded beams and cables, etc. 334 answered problems. 462pp. 5⅜ x 8½.   60754-2

MECHANICAL VIBRATIONS, J. P. Den Hartog. Classic textbook offers lucid explanations and illustrative models, applying theories of vibrations to a variety of practical industrial engineering problems. Numerous figures. 233 problems, solutions. Appendix. Index. Preface. 436pp. 5⅜ x 8½.   64785-4

STRENGTH OF MATERIALS, J. P. Den Hartog. Full, clear treatment of basic material (tension, torsion, bending, etc.) plus advanced material on engineering methods, applications. 350 answered problems. 323pp. 5⅜ x 8½.   60755-0

A HISTORY OF MECHANICS, René Dugas. Monumental study of mechanical principles from antiquity to quantum mechanics. Contributions of ancient Greeks, Galileo, Leonardo, Kepler, Lagrange, many others. 671pp. 5⅜ x 8½.   65632-2

Physics

OPTICAL RESONANCE AND TWO-LEVEL ATOMS, L. Allen and J. H. Eberly. Clear, comprehensive introduction to basic principles behind all quantum optical resonance phenomena. 53 illustrations. Preface. Index. 256pp. 5⅜ x 8½.   65533-4

ULTRASONIC ABSORPTION: An Introduction to the Theory of Sound Absorption and Dispersion in Gases, Liquids and Solids, A. B. Bhatia. Standard reference in the field provides a clear, systematically organized introductory review of fundamental concepts for advanced graduate students, research workers. Numerous diagrams. Bibliography. 440pp. 5⅜ x 8½.   64917-2

QUANTUM THEORY, David Bohm. This advanced undergraduate-level text presents the quantum theory in terms of qualitative and imaginative concepts, followed by specific applications worked out in mathematical detail. Preface. Index. 655pp. 5⅜ x 8½.   65969-0

ATOMIC PHYSICS (8th edition), Max Born. Nobel laureate's lucid treatment of kinetic theory of gases, elementary particles, nuclear atom, wave-corpuscles, atomic structure and spectral lines, much more. Over 40 appendices, bibliography. 495pp. 5⅜ x 8½.   65984-4

AN INTRODUCTION TO HAMILTONIAN OPTICS, H. A. Buchdahl. Detailed account of the Hamiltonian treatment of aberration theory in geometrical optics. Many classes of optical systems defined in terms of the symmetries they possess. Problems with detailed solutions. 1970 edition. xv + 360pp. 5⅜ x 8½.   67597-1

THIRTY YEARS THAT SHOOK PHYSICS: The Story of Quantum Theory, George Gamow. Lucid, accessible introduction to influential theory of energy and matter. Careful explanations of Dirac's anti-particles, Bohr's model of the atom, much more. 12 plates. Numerous drawings. 240pp. 5⅜ x 8½.   24895-X

ELECTRONIC STRUCTURE AND THE PROPERTIES OF SOLIDS: The Physics of the Chemical Bond, Walter A. Harrison. Innovative text offers basic understanding of the electronic structure of covalent and ionic solids, simple metals, transition metals and their compounds. Problems. 1980 edition. 582pp. 6⅛ x 9¼.   66021-4

HYDRODYNAMIC AND HYDROMAGNETIC STABILITY, S. Chandrasekhar. Lucid examination of the Rayleigh-Benard problem; clear coverage of the theory of instabilities causing convection. 704pp. 5⅜ x 8¼.   64071-X

INVESTIGATIONS ON THE THEORY OF THE BROWNIAN MOVEMENT, Albert Einstein. Five papers (1905–8) investigating dynamics of Brownian motion and evolving elementary theory. Notes by R. Fürth. 122pp. 5⅜ x 8½.   60304-0

THE PHYSICS OF WAVES, William C. Elmore and Mark A. Heald. Unique overview of classical wave theory. Acoustics, optics, electromagnetic radiation, more. Ideal as classroom text or for self-study. Problems. 477pp. 5⅜ x 8½.   64926-1

CATALOG OF DOVER BOOKS

METHODS OF THERMODYNAMICS, Howard Reiss. Outstanding text focuses on physical technique of thermodynamics, typical problem areas of understanding, and significance and use of thermodynamic potential. 1965 edition. 238pp. 5⅜ x 8½. 69445-3

TENSOR ANALYSIS FOR PHYSICISTS, J. A. Schouten. Concise exposition of the mathematical basis of tensor analysis, integrated with well-chosen physical examples of the theory. Exercises. Index. Bibliography. 289pp. 5⅜ x 8½. 65582-2

RELATIVITY IN ILLUSTRATIONS, Jacob T. Schwartz. Clear nontechnical treatment makes relativity more accessible than ever before. Over 60 drawings illustrate concepts more clearly than text alone. Only high school geometry needed. Bibliography. 128pp. 6⅛ x 9¼. 25965-X

THE ELECTROMAGNETIC FIELD, Albert Shadowitz. Comprehensive undergraduate text covers basics of electric and magnetic fields, builds up to electromagnetic theory. Also related topics, including relativity. Over 900 problems. 768pp. 5⅝ x 8¼. 65660-8

GREAT EXPERIMENTS IN PHYSICS: Firsthand Accounts from Galileo to Einstein, edited by Morris H. Shamos. 25 crucial discoveries: Newton's laws of motion, Chadwick's study of the neutron, Hertz on electromagnetic waves, more. Original accounts clearly annotated. 370pp. 5⅜ x 8½. 25346-5

RELATIVITY, THERMODYNAMICS AND COSMOLOGY, Richard C. Tolman. Landmark study extends thermodynamics to special, general relativity; also applications of relativistic mechanics, thermodynamics to cosmological models. 501pp. 5⅜ x 8½. 65383-8

LIGHT SCATTERING BY SMALL PARTICLES, H. C. van de Hulst. Comprehensive treatment including full range of useful approximation methods for researchers in chemistry, meteorology and astronomy. 44 illustrations. 470pp. 5⅜ x 8½. 64228-3

STATISTICAL PHYSICS, Gregory H. Wannier. Classic text combines thermodynamics, statistical mechanics and kinetic theory in one unified presentation of thermal physics. Problems with solutions. Bibliography. 532pp. 5⅜ x 8½. 65401-X